Walter Smith

Teachers' manual for freehand drawing in primary schools

Walter Smith

Teachers' manual for freehand drawing in primary schools

ISBN/EAN: 9783337282592

Printed in Europe, USA, Canada, Australia, Japan

Cover: Foto ©Paul-Georg Meister /pixelio.de

More available books at **www.hansebooks.com**

TEACHERS' MANUAL

FOR

FREEHAND DRAWING

IN PRIMARY SCHOOLS.

BY

Prof. WALTER SMITH,

State Director of Art Education for Massachusetts.

Intended to accompany the American Drawing-Cards, by the same author.

BOSTON:
JAMES R. OSGOOD AND COMPANY,
(LATE TICKNOR & FIELDS, AND FIELDS, OSGOOD, & CO.)
1875.

CONTENTS.

	PAGE.
INTRODUCTION	5
CENTRE, POINTS, AND STRAIGHT LINES	19
STRAIGHT LINE FIGURES .	40
SIMPLE CURVES . .	55
COMPOUND CURVES.—ELLIPSE AND OVAL	72
COMPOUND CURVES.—REVERSED CURVES, THE OGEE AND ABSTRACT CURVES	86
PRACTICAL DESIGN	105
PRACTICAL DESIGN.—LEAVES, BUDS, AND FLOWERS CONVENTIONALIZED . . .	132
PRACTICAL DESIGN.—FRUIT, COMPOUND LEAVES, AND SPRIGS CONVENTIONALIZED	150
FORMS OF LEAVES AND FLOWERS	159

INTRODUCTION.

THAT it is practicable to teach drawing in the public schools is no longer a matter of doubt; and that the study is one of great industrial and educational value, when properly taught, no person, who has carefully investigated the subject, questions for a moment. The drawing, however, must be wide in its range, rational, and systematic. It must be much more than simple copying: indeed, it must be, in the main, something wholly different from that; and it must be much more than picture-making, — even good original picture-making. Those who think otherwise sadly misapprehend the scope and value of one of the most practical studies.

It will be observed, as characteristic of this course of drawing, both in its primary and advanced grades, that the picture-element, as such, is almost entirely excluded. The author is well aware, that, in excluding many pretty things which have usually been considered proper for drawing exercises, he sacrifices something of superficial attractiveness; but he omits them because the best artistic, practical, or educational results cannot otherwise be produced. He excludes them from even this primary course, because he would here lay a sound basis for more advanced and severer work, when pupils reach the grammar schools. Hence, while the author is aware that pupils are anxious to begin what may be called "pretty work," as early as possible, he wishes teachers clearly to understand that picture-making, so far as the public schools are concerned, should always be regarded as a secondary matter. He also wishes teachers to remember that the correct representation of any object, as really seen, involves principles of perspective, with light and shade, which young pupils cannot understand; which cannot, indeed, be understood by any one without much hard study. Hence, to set children drawing from copies involving these features is to set them to reproducing, in a merely mechanical way, what they cannot possibly comprehend. Instead of leading learners into such ruts, the author desires, first of all, to make them acquainted with the beauties of pure form and with the principles of good design. Therefore, it is that the exercises in this primary course are principally such as deal with mere outline representation, based on geometrical forms, and illustrating leading principles of practical design. It is only when pupils have well mastered this stage that they are prepared to take up intelligently, and in a thorough manner, perspective, and model and object drawing.

DRAWING SHOULD BE TAUGHT BY REGULAR TEACHERS.

Drawing can, and should, be taught by the regular teacher. One who understands the general principles of teaching can teach drawing successfully, without any special artistic gifts. Thousands of persons teach

INTRODUCTION. 7

arithmetic successfully who possess no special mathematical gifts; and there is nothing about drawing so difficult to master as some of the features of mathematics. Let it, then, be accepted as true, that there is nothing about drawing to prevent any person of fair capacity becoming proficient enough to teach it intelligently and with success. It is clear, from the present movement in favor of art education, that the teaching of drawing will soon be required in all public schools; and it must, therefore, be taught by the regular teachers. Indeed, it is best, for various reasons, that it should thus be taught.

THE OBJECT OF THIS PRIMARY COURSE.

This Manual, with the cards which accompany it, is intended for a course of instruction in slate-drawing in primary schools. The exercises are so simple, and so gradually progressive, that teachers, though they may have had no previous instruction in drawing, can master them, if they choose, without assistance, and intelligently lead any class of young pupils who happen to be under their charge. There is no better way to teach one's self than teaching others. With the cards in the hands of the pupils, furnishing them good copies, it is not at all essential that teachers should be extraordinarily skilful in drawing on the blackboard. The Manual will acquaint them with all the principles involved: these principles they can explain orally to the pupils. It is the clear setting forth of these principles that constitutes the feature of chief importance in teaching drawing. The pupils, with their card copies before them, aided by the oral explanation, and by such illustrations (even though they be rude) as the teacher may give on the board, will make very satisfactory progress.

It is the leading object of these cards to teach proportion, the simple figures of plane geometry, the principles of practical design, and to familiarize the pupil with beautiful forms. To immediately follow these cards in the upper classes of primary schools, or in intermediate schools, the author has prepared three small books in free-hand drawing, to be used by pupils in beginning to draw on paper. The exercises in these books apply the principles of the instruction given in the card-exercises, and teach more fully the drawing in outline of objects with regular features.

REDUCTION AND ENLARGEMENT.

Teachers should make themselves sufficiently familiar with the exercises, by drawing them, to be able to point out their features readily and clearly. When working on the blackboard, they should proceed slowly, explaining each point as they go along, and requiring the whole class to work together, even though some are obliged to execute their drawings very rudely.

It is intended that the figures given in the first chapter of the Manual shall be drawn on the blackboard by the teacher to a scale of one foot for every half inch of the copy. Then the pupils are to copy them on their slates, making them of a definite size stated by the teacher. The size of the figures will be governed by the size of the pupils' slates. This will teach *reduction*. It is intended that the forms on

INTRODUCTION. 9

the cards shall be copied by the pupils from the cards; the teacher indicating the proportions, or scale, which should be much larger than in the copy. This will teach the pupils *enlargement*. The two features of practice, reduction and enlargement from copies, should go together: both are essential to good progress in drawing. If each pupil is supplied with card copies, the labor of the teacher will be greatly abridged; since all can be taught the same thing at the same time, while the advancement of the whole class will be more satisfactory than it can be otherwise. The figures given in the other chapters of the Manual, and not found on the cards, are intended for blackboard copies, to be drawn by the teacher as introductory exercises to the use of the cards, or for dictation exercises to accompany the cards.

GEOMETRICAL DEFINITIONS.

The definitions of plane geometry, illustrations of which are employed as a basis of freehand drawing, should be given to, and repeated by, the youngest pupils. They should be explained, not so much by words, as by freehand illustrations drawn on the blackboard as accurately as possible. Figures of this kind, when seen, children readily understand. The copies in the first chapter may be repeated as often as writing lessons; but, instead of continued repetitions of the same exercise, the whole series in the first chapter should be gone over once, and then be taken again from the first, with a better understanding and increased skill. When teaching children to draw, above all things avoid wearying them by repetition of uninteresting elementary work. They have no appreciation of abstract truths, nor of the words used to express such truths; but they are exceedingly sensitive to visible forms, and readily acquire a knowledge of them. It is possible to reach their understanding through the sense of sight sooner than by appeals to either of the other senses, or by logical explanations. Geometric accuracy in freehand or model drawing is not to be expected of any one: the standard by which the efforts of little children are judged should, therefore, be a very merciful one.

POWER OF OBSERVATION.

Since intelligent drawing is always an expression of knowledge, the power of observation must be cultivated from the outset. Pupils must be trained to discover the features of whatever they are to draw before they begin their work. To take a simple illustration: When the pupils have learned the names of the different straight lines and angles, the square may be given as an exercise. The teacher should draw it on the blackboard, requiring the pupils to name each line as drawn, and, when the figure is completed, to give the number and names of the angles it contains. After the diameters and diagonals have been drawn, the pupils should then be required to state the effect of these lines upon the original angles, and to give the number and names of the new angles and triangles. Other exercises may be treated in a similar manner.

CLASS ANALYSIS OF FORMS.

When the cards are used, a class analysis of the figure should frequently precede the drawing; that is, the pupils should be required to describe the general geometrical form of the figure, and the various lines which it contains. To take card exercise No. 33 for example: The analysis would show the pupils, before they began to draw, that an 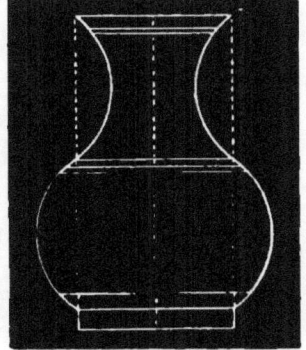 oblong would enclose the vase; that the vase is symmetrically arranged on an axis, its two sides being alike; that it is composed of compound curves and horizontal lines; that the top and bottom have the same width; that the widest part is near the bottom; that the narrowest part is the neck near the top; that it has a base; that the compound curves make both the widest and narrowest parts; that they do not reach to the bottom; that there is no ornament on the vase except the horizontal lines. It will be seen at once that this preliminary analysis must develop the observing powers, must firmly fix in the mind the principles of drawing and design, must enable the pupils to do their work intelligently.

First of all, then, the eye is to be educated to distinguish form, and the mind to comprehend principles: the correct manual rendering of the form is a simple matter, which will come by practice. In the early lessons, therefore, teachers should lay more stress upon developing observation and understanding than upon mere dexterity in execution. It is rare, almost accidental, for young pupils to draw clear unbroken lines; while lines varying in thickness must be expected from them for a long time. In criticising their efforts, form, irrespective of beauty of execution, should be mainly considered at first. When form has been fairly mastered, then teachers may begin to insist upon better lines and more careful workmanship. One thing should thus be taken at a time.

DRAWING ON THE BLACKBOARD.

Drawing on the blackboard should be practised at least once a week by all the pupils; a section working on the board every time a lesson is given, the others working at the same time on their slates. Pupils will be delighted with this, while it will afford them variety of practice, which is important. It will, in particular, allow them to make their drawings on a large scale. This work on the board may be termed the completest exemplification of freehand drawing.

LENGTH OF LESSONS.

Short lessons, often repeated, are better for young pupils than long ones at greater intervals. Four lessons a week, of half an hour each, or six of twenty minutes each, will be a fair amount of time for children under ten years to give to drawing. After that age the lessons may be longer and not so frequent, while new subjects may be taken up. When instruction is given at all in drawing, the lessons should be sufficiently frequent and sufficiently long to keep the pupils alive to their work, and thoroughly interested. Two short lessons a week will not do this. Four lessons will much more than double the results, and six lessons much more than treble them. This is not only true of drawing; but every teacher who has tried the experiment knows that it is also true of arithmetic or geography, for example. In neither of these studies would any one expect to make any satisfactory progress with only two short lessons a week. Indeed, the time would be regarded as, perhaps, worse than wasted; since the little progress actually made would not compensate for the disgust many of the children would acquire for the study thus treated. If, then, drawing is to be taught at all, it should be taught in earnest, and not as a merely tolerated study. Even its decided attractiveness will not fully compensate for the lack of earnest work.

Again, if any new study, whatever, is to be put into a school, it should receive for a while, if possible, special attention, in order to emphasize its importance in the minds of the pupils, and to make a strong beginning. To begin well is half the battle. When the new study is thus once firmly established, it may then take its place with the other regular studies. Thus should the schools welcome drawing, which is destined to contribute so much towards the future prosperity and culture of the country.

DICTATION EXERCISES.

It is remarkable how few persons can accurately describe the true form, size, proportions, color, and character of any thing which the eye sees. People have general impressions, which, brought to the test of description, are found to be vague and altogether unreliable. Drawing, preceded by criticism and analysis of copies or objects, in which well-chosen terms describe details, cannot fail to educate even the reasoning powers as well as the imitative and artistic. For this purpose, lessons should also be frequently dictated by the teacher, without illustration in books, on cards, or on the blackboard.

For this kind of instruction, artificial objects, having geometric forms, are better suited than natural objects. For example, houses of simple proportions may thus be dictated and drawn. When such a drawing has been executed, it is then advisable to require the pupils capable of doing so to write a minute description of it, giving its form, size, proportions, &c. There could not possibly be a better subject for a composition, when it is desired to teach exactitude in the use of language.

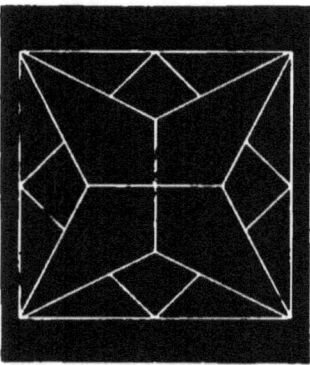

A simple illustration of the dictation exercise is here given. The drawing of the annexed geometrical form may be dictated thus: Draw a square, its diameters and diagonals. Halve the semi-diameters. From each corner of the square, draw a straight line to the nearest point of division on each diameter. On the diameters of the first square, as diagonals, draw a second square. Erase the diagonals of the first square, and the outer divisions of its diameters; also those parts of the inner square which go behind the points of the star. If the pupils heed what is dictated, they will produce the required figure; if not, they will draw something else. After the pupils have finished their work, the teacher will draw the figure correctly on the blackboard.

The object, then, of dictation exercises is to show that forms and words are interchangeable, both being mediums for the expression of thought; to teach pupils to use either language with equal facility; to translate and retranslate from one into the other. Such exercises, beginning with simple geometric forms, may be continued, until the most elaborate irregular forms can be drawn from dictation, if the language describing them be accurate and terse. Artisans have constant use for the power which is acquired from dictation exercises.

MEMORY EXERCISES.

Pupils should be frequently required to produce, from memory, exercises which they have previously drawn. This will show the teacher how much of previous instruction is retained, while it is valuable simply as a means for developing the power of memory. Therefore, as soon as pupils have got a limited knowledge of forms, they should begin to reproduce these forms without seeing the copies and within a specified time. It is also urged that the pupils, leaving their cards at school, be occasionally required to do their drawing exercises at home, from memory, and, putting them into a little book provided for the purpose, bring them to the teacher for inspection.

With each series of cards there is a five-inch scale, the inches divided into halves, quarters, and sixteenths. This scale should be used only for correcting lines after they have been drawn, and not to help in first drawing the lines. Teachers must see that pupils do not use the scale in the manner prohibited. It is essential that the first lines of any drawing should be approximately correct, otherwise the remaining lines will be wholly out of place. Hence, when the first proportional lines have been drawn as nearly correct as possible by the eye alone, pupils should be permitted to measure them, and to correct them if they need correction. It is to be distinctly understood, however, that the measuring is not to be permitted until the lines have been drawn.

Some teachers say, "Never draw a straight line freehand if you have the means of ruling it;" yet it is better thus to draw straight lines than to rule them, unless it is absolutely essential, for some special reason, that they be mathematically straight, both because it is good practice for the eye and hand, and makes the pupil independent of rulers. The same may be said of drawing circles. Of course, when any thing of importance depends upon a circle being perfectly drawn, it is best to use a pair of compasses; but a circle practically exact may be drawn freehand with very little experience, and it is educational thus to draw it. In practice, it will be found more difficult to rule either vertical or horizontal straight lines, than to draw them by the aid of the eye alone. A somewhat extended experience in teaching, with continued observation of the work of many pupils, has convinced the author that those who draw vertical and horizontal lines freehand, besides doing their work in less time, generally make their lines more nearly upright and level than those who rule them. Those only draw well who draw intelligently.

POSITION.

When vertical and horizontal lines are drawn, the slate or paper should be kept in one position in front of the pupil, its edge parallel with the edge of the desk. When oblique lines are drawn, the slate or paper may be kept in the same position, while the hand is turned so that the vision may not be obstructed. When curves are drawn, it will be best to allow the slate or paper to be put in any position the pupil may prefer, in order that the hand may form the centre of the curve drawn, which can be, in some cases, only when the drawing is turned sidewise or upside down.

PENCILS AND MODELS.

The slate pencils should be soft, long, well-pointed, and held about an inch and a half from the end, far enough to keep the fingers from obstructing the vision.

For the purpose of illustrating many of the exercises, and accelerating the advance of the pupils, every primary school should have a box of models, which can be procured of the publishers of this book. Such a box would contain a sphere, cube, oblong block, triangular prism, hexagonal prism, cone, cylinder, rectangular pyramid, square frame, cross, and some other pieces.

In conclusion, the author would say that the child who draws from five years of age to fifteen ought to be able to draw any created thing, and will be able to do so when we teach well what every child ought to know. There is no royal road to drawing; and those who profess to offer such a road to us may be suspected of a desire to mislead with a regal name which represents nothing.

CHAPTER I.

CENTRE, POINTS, AND STRAIGHT LINES.

WHEN young children begin to draw, teach them, first, the meaning of the word "centre." Require them, by way of illustration, to touch the centre of their slates with the end of their pencils. This centre we will call a point.

A Point.

A point is simply position or place, without size, and may be indicated in different ways.

Explain, next, the meaning of the words "right" and "left." Draw a vertical line on the blackboard; to the right of this line draw another, explaining that the last is to the right hand of the first. In the same way illustrate the meaning of the word "left." Next, draw a horizontal line, and in a similar manner explain the meaning of the words "above," "below."

Always make certain, in some way, that children have a clear understanding of whatever is used for the purpose of further extending their knowledge. It is well to take nothing for granted, but to explain the simplest things. When the celebrated Faraday lectured before children at the Royal Institute, London, he was accustomed to illustrate the most common phenomena. He would not take it for granted that the children knew that an apple, when unsupported, will fall to the ground; but he would actually illustrate the fact.

STRAIGHT LINES.

A VERTICAL LINE. — *A vertical line is a straight line which runs up and down, inclining neither to the right nor to the left, neither backwards nor forwards* (a).

A HORIZONTAL LINE. — *A horizontal line is a level straight line, inclining neither up nor down* (b).

AN OBLIQUE LINE. — *An oblique line is a straight line, neither vertical nor horizontal, but slanting* (c).

First, draw the three kinds of lines on the board, and give their names. Then require the pupils to draw them on their slates, without turning their slates. Each line should be drawn many times, until the pupils can instantly distinguish between the different lines; until they can instantly name them when drawn by the teacher, or instantly draw them when named by the teacher. A straight stick held in different positions will serve nicely to illustrate the meaning of the words "horizontal," "vertical," "oblique." Request the pupils to name objects in these different positions.

PARALLEL LINES. — *Parallel lines are lines running side by side in the same direction, and are always at the same distance from one another throughout their whole length. They may be straight or curved, vertical, oblique, or horizontal.*

Draw on the board parallel vertical (A), horizontal (B), and oblique lines (C), and require the pupils to repeat them on their slates, without turning their slates. This is an exercise to which you should frequently revert. When the 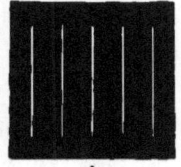 pupils have drawn two lines parallel, require them to draw a third in the centre between the two, and thus let them continue to halve the spaces between the lines, until the lines are very close together.  The pencil should have a fine point for this work, and no two lines should be allowed to touch each other. This is excellent practice. As soon as the pupils have acquired some skill in judging distances, require them to draw parallel lines of definite length, as one inch, two inches, &c.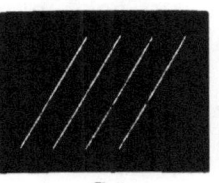

Pupils should have a clear understanding of parallel lines. There are two things requisite to make lines parallel: they must lie side by side, and they must have the same direction. When, therefore, we want two parallel vertical, or two parallel horizontal lines, it is not enough to say, "Draw two vertical lines," or, "Draw two horizontal lines;" for in neither case need the lines be drawn side by side. Nor is it enough when two parallel lines are wanted, simply to draw them side by side; for side by side does not necessarily mean that the lines must have exactly the same direction. Hence we must say parallel vertical lines, parallel perpendicular lines, parallel oblique lines, or parallel horizontal lines, as the case may be, if we would say just what we mean. As further progress is made in this course of drawing, it will be more clearly seen why so much stress is laid upon the exact use of terms.

See that the pupils always have long and well-pointed pencils: they should never be allowed to draw with stubs or blunt pencils. To make sure that the pencils are always in good condition, keep them in a box under your own charge, to be distributed and taken up at every lesson. Appoint some one of the pupils, from day to day, to sharpen the pencils, doing the work where it will not disturb the school. Without good material, it is impossible to draw rapidly and well.

The pencil should be held about an inch and a half from the point, and so held that the hand will not come between the eye and the space where the line is to be drawn, obstructing the vision.

When drawing straight lines in different directions, as horizontal, vertical, and oblique, instead of turning the slate or paper, turn the hand. Observe these two things: always keep the hand and fingers in a position of freedom; never let the hand obstruct the sight.

Draw the shortest lines with the fingers alone; those somewhat longer draw with the hand, using the wrist joint; the next grade draw with the fore-arm, using the elbow joint; those longer yet, which will usually be on the blackboard, draw with the whole arm, using the shoulder joint. Occasionally, in drawing a very long horizontal line on the blackboard, it

STRAIGHT LINES. 23

is necessary to carry the body forward as the line is drawn. It is best, when drawing a vertical line on the blackboard, to stand directly in front of where the line is to be drawn, with the side partially turned towards the board, and far enough away to let the hand, when the arm is extended, fall easily carrying the crayon down. Never draw a line with a jerky movement, but always with a uniform motion, and slow enough for the eye to follow.

DIVIDING STRAIGHT LINES AND JUDGING DISTANCES.

Draw on the blackboard a vertical line (1), and divide it into two equal parts. Explain that the dividing mark is in the centre of the line. Draw another line (2), and divide it into four equal parts; first, by dividing it into two equal parts, and then each of the halves into two other equal parts. Draw 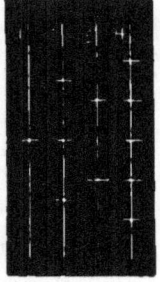 another line (3), and divide it into three equal parts. Dividing a line into three equal parts is more difficult than dividing it into two equal parts. Draw another line (4), and divide it into six equal parts; first, by dividing it into two equal parts, and then each of the halves into three equal parts. While the teacher draws these lines on the blackboard, the pupils should draw lines of a given length on their slates, and divide them into halves, quarters, thirds, and sixths, following the teacher. Exactness in the division of lines should not be expected on the part of young pupils at first. It is sufficient, if they get the idea.

Accuracy will come by practice. In these early stages, permit the pupils to correct their faults in free measurement by the aid of the scale which accompanies each series of cards. This practice will be found the surest way of educating them to correct observation. Before resorting, however, to the use of the scale, the pupils should do their best with their eyes alone. Drawing and dividing lines in this manner should be practised in connection with other lessons, until pupils acquire a considerable degree of skill in determining definite distances and proportions.

Young children, unless they have received positive instruction, seldom have any clear idea of the length of an inch, a foot, a yard, or mile, though they may have heard the words used hundreds of times. If questioned, they frequently give answers that astonish the questioner.

Suppose, further, that it is required to draw a line, but of no particular length, and to divide it into five, seven, eleven, thirteen, or any other prime number of equal parts. In that case, having drawn a line, divide it into four equal parts, and then add one part, to make five; or divide it into eight equal parts, and then erase one, leaving seven, if that is the number required. That is, first draw a line, and divide it into a composite number of parts, then add or erase, as may be required. This is not, of course, a suitable exercise for the youngest pupils.

Frequently ask the pupils how they would proceed to divide a line into any designated number of equal parts. If, for example, the number is six, the answer should be: Divide the line into two equal parts;

then divide each part into three equal parts, giving six.

COMBINING STRAIGHT LINES. — ANGLES.

D.

As soon as the pupils have acquired a knowledge of straight lines, they should begin to combine them. Draw on the blackboard a vertical line of definite length, say twelve inches, and then require the pupils to draw a similar line, say four inches long, on their slates.

At this stage, the lines should all be drawn of definite lengths. Having first been drawn by the aid of the eye alone, they should afterwards all be corrected by the scale. Such practice is the beginning of the study of proportion.

Having drawn the vertical line, next draw a horizontal line of the same length; draw it from the lower end and to the right of the vertical line. The two lines, thus united, form what children will call a corner, but what we will call an angle. This particular angle (D) is called a right angle, not, however, because it opens to the right.

Right Angles.

A right angle is formed by the meeting of two straight, or right, lines perpendicular to each other.

The pupils should be taught to distinguish clearly between a vertical line and a perpendicular line. A vertical line is always perpendicular; but a perpendicular line is not always vertical. One line is said to be perpendicular to another line when the two meet so as to form a square corner, or right angle;

yet both lines, when considered with reference to their position, may be oblique. But one line is never said to be vertical to another. When a right angle has been formed on a slate by the meeting of a vertical line and a horizontal line, it will remain a right angle, though the slate be partly turned around, and the lines become oblique. When a knife is half open, it forms a right angle: the blade is perpendicular to the handle, in whatever position the knife may be held, while the handle is perpendicular to the blade. Frequently illustrate this important distinction between the words vertical and perpendicular, and always insist upon the pupils using the words correctly.

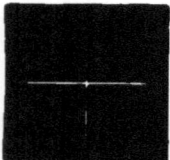
E.

Now extend the horizontal line in D the same distance to the left of the vertical line already drawn, and explain that the corner thus formed on the left is also a right angle. By next extending the vertical line the same distance below the horizontal line, two more right angles will be formed; and you will have four right angles meeting in a point, or at the centre, as E.

The size of an angle does not depend at all on the length of the lines forming it; for

An angle is the difference in the direction of two lines which meet, or would meet if extended far enough.

Therefore, the four angles meeting in E would be no larger, though the lines were drawn longer.

Draw right angles on the board in various ways. Thus, draw a horizontal line; from its left end, draw a vertical line downwards. Again, draw a horizontal

STRAIGHT LINES.

line; from its right end, draw a vertical line downwards; also, from above, draw a vertical line to join the right end of the horizontal line. Again, draw an oblique line; and, perpendicular to its centre, draw another which will make two right angles on the first oblique line, as shown at F. Pupils repeat on their slates.

Acute Angles.   **F.**

Any angle less than a right angle is an acute angle.

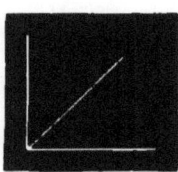

Draw a right angle, and, midway between the two lines, draw an oblique line meeting the first two in the corner, or vertex, of the angle at G. This oblique line divides the right angle, and makes two sharper corners of it. These sharper corners are called acute angles, because they are sharper, or more pointed, than a right angle.

G.

Obtuse Angles.

An angle greater than a right angle is called an obtuse angle.

This figure at H shows what is meant by "greater than a right angle."

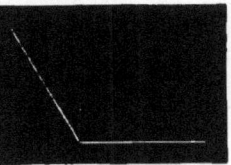

H.

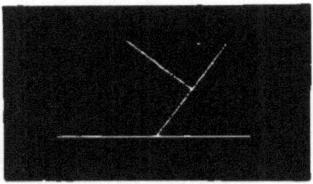

I.

Draw a horizontal line as in I; then an oblique line, joining the horizontal line at its centre; and then a third line, perpendicular to the second line. The union of the horizontal and oblique lines forms one acute angle and one obtuse angle; the former to the right of the oblique line, the latter to the left. The union of the two oblique lines forms two right angles. Thus draw the different angles in various positions, and require the pupils to name the angles, until they can readily recognize them, however they may be drawn. Also require the pupils to make the different angles, by placing two sticks in different positions. Explain clearly that an angle is the *difference in direction* of two lines, and not the lines themselves, nor the space between them. An angle is formed when two lines tend towards each other, though they do not extend far enough to meet.

Good drawing models are the best objects with which to illustrate the principles of geometrical drawing.

The following letters drawn with single lines illustrate the different angles. Letters E, F, H, L, contain right angles. Letters M, N, V, W, Z, contain acute angles. Letters A, K, X, Y, contain acute and obtuse angles.

COMBINING STRAIGHT LINES. — TRIANGLES.

A triangle is a figure having three straight sides, enclosing three angles.

A Right-angled Triangle.

A right-angled triangle has one of its angles a right angle (J).

Draw a right angle, by joining the end of a vertical line to the end of a horizontal line. Draw an oblique line, joining the other ends of the lines. These three lines will give a three-sided figure, as at J, which is called a triangle, because it has three angles; and a right-angled triangle, because one of the angles is a right angle.

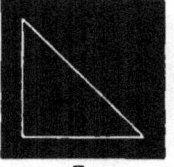

J.

Here is a curious fact about triangles: The three angles are always equal to two right angles. If one is a right angle, as in the present instance, then the other two will be equal to a right angle. This may be easily illustrated with paper. Cut paper into triangles of various shapes; then cut off the corners of any one of the triangles, and lay them side by side, so that the points will be close together, when they will be found exactly to fill the space of two right angles. Mark the corners before they are cut off; otherwise, the right corners may not be laid together.

An Equilateral Triangle.

An equilateral triangle has three equal sides and three equal angles (K).

It will require considerable practice to draw an equilateral triangle correctly. For the present, it will be sufficient if the pupils get the idea.

K.

They should name the lines forming the triangles, also the kinds of angles formed by the lines. A triangle is the strongest of all figures. The position of three sticks thus united cannot be changed without breaking one of the sticks.

An Isosceles Triangle.

An Isosceles triangle has two only of its sides equal (L).

L.

Draw a horizontal line, and mark its centre. From this centre, draw a vertical line upwards. Join the end of the vertical line to the ends of the horizontal line by two oblique lines, as in L. Erase the vertical line, when an isosceles triangle will remain. The word isosceles means having equal legs, as this triangle has two equal sides. An equilateral triangle is an isosceles triangle; but all isosceles triangles are not equilateral.

A Scalene Triangle.

M.

A scalene triangle has no two of its sides equal.

Draw three straight lines of unequal length, uniting their ends, as at M. The word scalene means limping. This triangle can be placed on no one of its sides without leaning to the right or left.

COMBINING STRAIGHT LINES. — RECTANGULAR FIGURES.

Any figure having four straight sides and four right angles is called a rectangle.

A Square.

A square is a figure having four equal sides and four right angles.

STRAIGHT LINES.

Draw a right angle opening to the right, with the horizontal and vertical lines of the same length. From the upper end of the vertical line draw towards the right a horizontal line of the same length as the one below. Next, draw a vertical line connecting the right ends of the two horizontal lines. Thus we have a square (N).

Let the pupils frequently name the lines, state if any are parallel, and also give the number and kind of angles the square contains.

The pupil having learned to draw a square, it will now be well for him to work within one of definite size.

DIAMETERS OF A SQUARE. — *A diameter of a square divides the square and its opposite sides into two equal parts.*

Draw a square as before. Divide each of its sides into two equal parts; connect the points of division by a vertical and a horizontal line, as in O. These lines give the diameters of the square, and, at the same time, divide it into four equal smaller squares.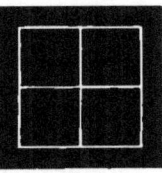

Let the pupil name the kind and number of angles in these four squares.

DIAGONALS OF A SQUARE. — *Lines connecting the opposite angles of a square are called the diagonals of the square.*

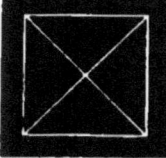

Having drawn a square as before, draw oblique lines connecting the opposite angles, as in P.

Add the diameters, and require the pupils to tell how many angles and triangles are now contained in the square, and to give the name of each.

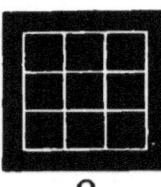

Having drawn a square as before, divide its horizontal sides into three equal parts, and then draw vertical lines connecting the points of division as in Q. Next, divide the vertical sides of the square into three equal parts, and then draw horizontal lines connecting the points of division. The square will thus be divided into nine smaller squares. It will now be well for the pupils to divide the square by a larger number of vertical, horizontal, and oblique lines. The oblique lines should be drawn parallel with the diagonals, while the other lines will be parallel with the sides of the square.

A Square constructed on its Diameters.

Draw two lines of equal length, one horizontal and the other vertical, and crossing at their centres. Through the ends of these lines, draw other lines of the same length, horizontal and vertical, with their ends uniting: the result will be a square, as in O.

Pupils should be required to construct squares in the manner first indicated, and also on their diameters, until they are familiar with both methods.

STRAIGHT LINES. 33

A Square constructed on its Diagonals.

Draw two lines of equal length, one horizontal, the other vertical, and crossing at their centres. Draw other lines connecting the ends of the first two lines, and the result will be a square, as in R. It will be well for the pupils to practise drawing squares in different positions.

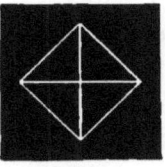

R.

An Oblong.

An oblong has four right angles, but only its opposite sides are of equal length.

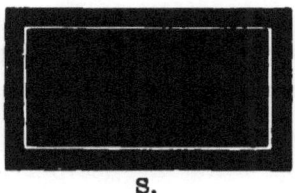

S.

Draw a horizontal line; from its ends, draw downwards two vertical lines, of equal length, but less than the horizontal line; connect the lower ends of the vertical lines by a second horizontal line, and the result will be an oblong, as in S. Frequently require the pupil to point out the difference between the oblong and the square.

COMBINING STRAIGHT LINES. — OBLIQUE FIGURES.

A Rhombus.

A Rhombus has four equal sides, but its angles are not right angles (T).

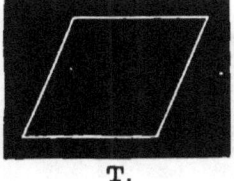

T.

It will be observed, that in the present instance, while two of the sides are horizontal, the other two are oblique. Of course the rhombus, like all other geometrical figures, may be drawn in any position, without change of character. It is not necessary that any of its sides should be horizontal, as it is not necessary that any of the sides of a square should be. The pupils should be required to contrast the rhombus with the square, and point out the difference between the two figures. Thus the square has four equal sides. So has the rhombus. The square has four right angles; but the rhombus has two acute and two obtuse angles. In defining a rhombus or a square, it is necessary, therefore, to describe both the sides and the angles, or the definition which is intended for the one may be taken for the other.

All figures having four straight sides are called quadrilateral, a term that embraces much more than rectangular.

By comparing and contrasting different geometrical figures, the pupils will learn as readily to distinguish and describe three or four, as they will learn one alone. This involves a valuable principle in teaching.

The pupils having drawn a rhombus, let them divide it into any number of similar figures, as they divided the square.

A Rhomboid.

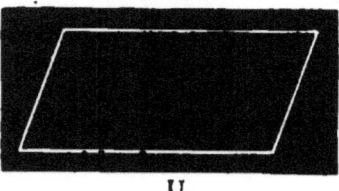

U.

A Rhomboid has four sides: none of its angles are right angles, and only the opposite sides are equal (U).

The termination *o-i-d* means like; hence rhomboid means like, or resembling, a rhombus. Require the pupils to contrast the rhomboid with the rhombus, also with the oblong, and point out the resemblances and the differences among the three figures. The pupils, having drawn a rhomboid, should be required to divide it into any number of similar figures. It will be well to do this repeatedly. The rhomboid and the oblong are both called parallelograms, because they have their opposite sides equal and parallel.

The teacher who is anxious to impress on the minds of the pupils as firmly as possible the different forms of plane geometrical figures, and to give additional variety to the instruction, will find a cord, with its two ends tied together, of service, when used in the following manner: Let three pupils take hold of the cord, each with one hand, and then, as requested by the teacher or any one of the class, let them so hold the cord as to form a right-angled triangle, an equilateral triangle, or any other triangle, as the scalene or isosceles. Next, let four pupils take hold of the cord, and, in a similar manner, form different four-sided figures. Five pupils could form the pentagon, six pupils the hexagon, and so on. All the figures formed by the same cord would have the same distance around them, and so would be isoperimetrical.

The preceding exercises should be fairly understood by the pupils before they advance farther. It is not necessary that they should be able to draw the figures accurately; but they should be able to give their names, and to describe them with readiness. Teachers can increase the love of their pupils for their work by drawing for them the simple outlines of objects which come under their daily observation in the schoolroom or in the vicinity. Simple forms, composed of straight lines, and exhibiting the geometric features of the preceding exercises, will occur to every teacher. Teachers should remember, that at this stage of the study, when examples are given on the board, they should be constructed with definite, simple proportions. Drawing should always be taught as *an expression of knowledge, and not simply as a copying process.* Hence pupils must be taught first to see and understand the proportions and properties of forms and objects, otherwise they can never draw them intelligently; and the earlier exercises should, therefore, be of such a character that they can comprehend them.

Every primary schoolroom should be supplied with a set of primary school models. These models are very useful for fixing in the minds of young pupils a knowledge of geometric forms.

This book is specially designed for the use of teachers, to show them how to execute the drawings, and how to instruct others. They are expected to explain every thing in language even more familiar than that of the book, that young children may be able clearly to understand the things taught. They should not, however, fall into the error of supposing that children, even quite young children, cannot be readily made to understand the meaning of long words, or scientific words, because they are long or scientific. Whenever the meaning of such words, for example, as perpendicular, equilateral, rhomboid, is visibly illustrated, children can understand the words as readily as they can understand the meaning of right, left, square, and will take pride in using them. Hence teachers should see that pupils are made acquainted with the exact force of the technical words that must be frequently used in teaching drawing, if those who teach would express themselves briefly and clearly, and have their pupils learn to do the same.

Pupils should draw frequently on the blackboard, a thing they are always delighted to do. Whether they draw on the blackboard or on the slate, they should not at first spend much time in an effort to make their lines beautiful. They should rather learn to draw with boldness and rapidity, even from the outset: accuracy of finish will come from practice.

It has been held by many that children and all others, when they begin to draw, should begin with natural objects, and not with severe geometrical forms. But careful examination of results in Europe, during the last twenty years, shows that the former is not the best way to begin, though it may be the best way to end; and especially it is not the best way to begin when the aim is to produce industrial results. Drawing from nature and from the human figure alone, or mainly, is a round-about and laborious way to such results. Of course a limited amount of drawing from nature, for the sake of variety, is not only to be permitted, but encouraged, at all times. The pupils then, even at the outset, should be made acquainted with geometrical forms and terms, that they may afterwards be able, understandingly, to describe and draw objects of all kinds; for objects, whether natural or artificial, are usually based on regular geometrical forms. Without a knowledge of these geometrical forms the artisan, in particular, can have no clear appreciation of his work.

Again, it is contrary both to experience and to the principles of good teaching to suppose that pupils can draw, with any degree of accuracy, a compound curve, like the ogee, or the delicately-waving lines which are seen in the forms of natural objects, when they are unable to draw a straight line; or to suppose they can draw, with any degree of accuracy, such irregular forms as those of animals and trees, whose proportions can only be guessed, when they cannot draw regular, symmetrical forms, whose exact proportions are known; or that they can draw objects from nature, involving the difficulties of shade and perspective, when they cannot draw from flat copies, without shade or perspective, those symmetrical forms which always have been, and still are, so extensively employed in practical design.

Hence it is that this course of drawing begins with geometrical forms illustrated by models; gives precedence to that which has definite proportions, and so permits the drawings to be verified by actual measurement; puts the symmetrical before the unsymmetrical; and works with flat copies, making the pupils familiar with pure form, and teaching the principles of practical design, before taking natural objects, which involve perspective, light, and shadow.

QUESTIONS. — What is a point? Describe the different kinds of straight lines. What is the difference between a vertical and a perpendicular line? What two things are requisite to make lines parallel? What is said of drawing lines in different directions? Of different lengths? How would you divide a line into six equal parts? Into seven? What is an angle? Describe the different kinds of angles? What is a triangle? Describe the different kinds of triangles? What is the difference between an isosceles triangle and an equilateral triangle? Between a quadrilateral and a rectangle? Between a square and a rhombus? Between an oblong and a rhomboid? What is the diameter of a square? Its diagonal? How should drawing always be taught? What is said of giving children long technical words? How does this course of drawing begin? Why does it so begin?

CHAPTER II.

STRAIGHT-LINE FIGURES.

THE pupils, having become familiar with the preceding exercises, should begin to draw from cards, under the direction of the teacher. Thus far they have been taught *reduction*, since they were required to make their drawings smaller than the teacher's blackboard copies. The card exercises will teach them *enlargement*, since they should be required to make their drawings much larger than the card copies. These two modes of practice, *by reduction and by enlargement*, should go hand in hand. Both are essential to good progress.

CARD-EXERCISE I.

Vertical, Horizontal, Oblique, and Parallel Lines.

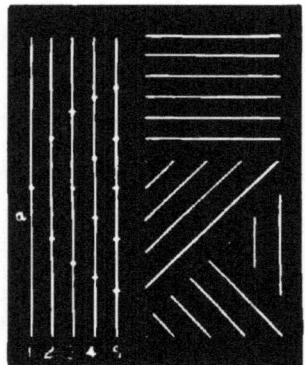

Draw vertical lines the whole length of the slate; then, beginning at one end, divide them into equal parts, first of indefinite length, next of definite length, as one inch, an inch and a half, &c. Draw horizontal and oblique lines, dividing them in a similar manner. Afterwards draw vertical, horizontal, and oblique lines, and divide them into a given number of equal parts, as two, three, four, six, eight, ten. To obtain four equal parts, first divide the line into two equal parts, and then divide each of these parts into two equal parts. To obtain six equal parts, first divide the line into two equal parts, and then each part into three equal parts.

Frequently, before the pupils begin to draw from a card, the teacher should require them, as a class and under his direction, to analyze the copy, and get a clear understanding of its features. Thus to analyze each copy would consume too much time, nor is such an analysis of each copy essential to a clear understanding of the exercises, as each feature is usually often repeated in the different exercises; but the preliminary analysis should receive quite frequent attention. Many elaborate drawings become easy of execution the moment analysis has made known their elementary lines or forms. Again, this analyzing process shows how much of the previous instruction has been retained, while it also tends to develop the perceptive powers of the pupils.

CARD-EXERCISE II.

Right-Line Figures.

Having placed a card in the hands of each member of the class, conduct the analyzing exercises after the following manner: —

FORM *a*.
Teacher. — What lines in this figure?
Answer. — Vertical, horizontal, and oblique lines.
Teacher. — Into how many parts is the vertical line divided?

42 TEACHERS' MANUAL.

Answer. — Four.

Teacher. — How would you begin to draw this figure?

Answer. — First draw the vertical line, and divide it

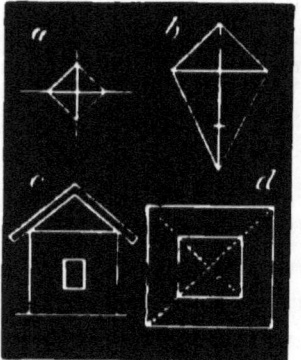

into two equal parts. Through the point of division, draw a horizontal line of the same length as the vertical line, one-half to the right, one-half to the left, of the vertical line.

Teacher. — What is the name of the figure drawn on the vertical and horizontal lines?

Answer. — A square.

Teacher. — How is it drawn?

Answer. — On its diagonals.

Teacher. — What is the diagonal of a square?

Answer. — A straight line connecting opposite angles of a square, and dividing the square into two equal right-angled triangles.

Teacher. — How many diagonals can a square have?

Answer. — Two.

Teacher. — When drawn, how do they divide the square?

Answer. — Into four equal right-angled triangles.

Teacher. — How do you obtain the points for this square?

Answer. — By dividing each half of the vertical and horizontal lines into two equal parts.

Teacher. — How do you then complete the square?

Answer. — By drawing oblique lines to connect the points of division.

It is not to be expected that young children will give answers correctly as here set down. The foregoing is presented only as a suggestive mode of conducting the analyzing exercise. Adapt the mode of analysis to the capacities of your pupils.

It will be observed that the answers are abbreviated in the text. Experience, however, has shown that it is well frequently to require small children, when answering questions, to make each answer a complete statement. Do this at times in the analysis of forms; Thus: —

Teacher. — What is the diagonal of a square?
Pupil. — The diagonal of a square is a straight line connecting opposite angles of the square, and dividing it into two equal right-angled triangles.
Teacher. — How many diagonals can a square have?
Pupil. — A square can have two diagonals.

Much importance is justly attached to this mode of answering, especially by small children. It is not only a better test of their knowledge, but trains them to clear statement. To practise it constantly, however, would become wearisome, and would consume too much time.

FORM *b.* — Draw a vertical line, divide it into thirds; through the upper point of division, draw a horizontal line equal to two-thirds of the vertical line; and then complete the figure by drawing the oblique lines.

FORM *c.* — Draw a square. Find the middle of the upper side, and, from this point, draw a vertical line upwards, equal to one-half of the length of the side of the square. From the upper end of the vertical line, draw oblique lines downwards to the left and right, striking the upper corners of the square, and projecting slightly beyond. Draw the parallel oblique lines, and finish as in the copy.

FORM *d*. — Draw a square and its diagonals. Halve each semi-diagonal, and through the points of division draw the inner square. Erase the dotted lines. These two squares are called concentric squares, because they have the same centre.

When the pupils are drawing from the cards, and after the first line to be drawn has been selected, state how long the line is to be made on the slates. When they have drawn the line of the designated length, and in the position it has on the card, then the teacher should draw it on the blackboard. Thus proceed with each line. In this way the teacher's drawing will confirm the work of the class, and not serve as a copy. In this way, too, the members of the class will be kept together, and instructed as one pupil. The aim should be to work with a good degree of rapidity, though the figures may not be accurately executed at first.

At this stage pupils should begin to draw both from memory and dictation. That is memory drawing, when the pupils are required to reproduce, without definite instructions, forms which they have previously drawn, as they might, in this case, be required to draw from memory an equilateral triangle, a square on its diameters, an oblong, a rhombus, or a rhomboid. The teacher would simply direct the drawing of any one of these figures, and the pupils would execute it from memory. That is dictation drawing, when the teacher dictates, but does not draw, the lines for the composition of a new figure. The

STRAIGHT-LINE FIGURES.

pupils draw, following the words of the teacher. Take figure A, for example.

The teacher dictates thus, while the pupils execute: "Draw a vertical line four inches in length; mark its centre, and from it, to the right, draw a horizontal line two inches in length. From the end of this horizontal line, draw downwards a vertical line two inches in length."

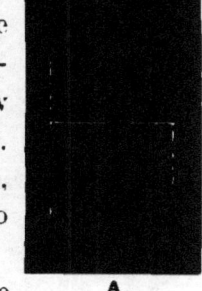

A.

This illustrates the general mode of conducting dictation exercises. The teacher must be very particular about the words used.

The memory exercises show whether the forms, and the modes of drawing them, have been clearly and permanently fixed in the minds of the pupils. Of course the memory is greatly strengthened thereby. The dictation exercises show the perfect interchangeability of words with lines and forms: that a form can be drawn from words as well as from a copy. They also strengthen the imagination; since the pupils are compelled to make a mental picture of what is required by the words before they can intelligently execute. Workmen are constantly obliged to work thus under dictation, — to interpret words by forms, and to use the imagination.

CARD-EXERCISE III.

Squares on Diameters and Diagonals. — Zig-zag Moulding.

Draw a square on its diameters. Using the diameters of the first square as the diagonals of a second square, draw the second inside the first.

This gives two concentric squares. Having divided

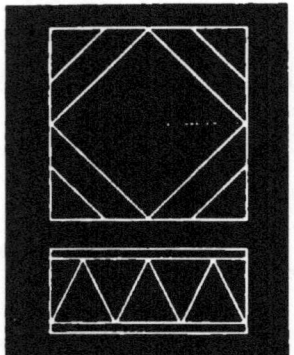

the sides of the outer square into four parts, connect the points of division by oblique lines parallel with the sides of the inner square. This is a good figure for the pupils to analyze before they begin to draw it.

ZIG-ZAG MOULDING.—Draw two parallel horizontal lines, of sufficient length to allow three or more squares to be drawn between them. Divide the upper side of each square into two equal parts, and, from the points of division, draw the oblique lines. Add a line above and a line below, parallel with the sides of the squares.

CARD-EXERCISE IV.

A Picture-Frame.

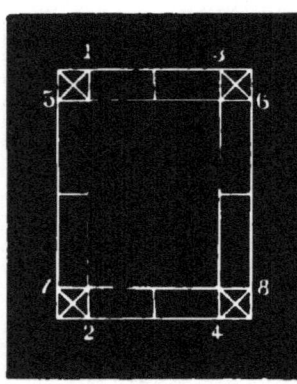

Draw an oblong, with its width equal to two-thirds of its height. Divide its width into six equal parts; draw the vertical lines 1 2 and 3 4; and then the horizontal lines 5 6 and 7 8, forming squares at the corners of the oblong. Draw the diagonals of the small squares; then add the short vertical and horizontal lines, which divide each side of the frame into two equal parts.

Card-Exercise V.

Oblong House.

Draw an oblong equal to two equal squares. To do this, first draw a vertical line of the required length, — two inches, for example; then draw from each end a horizontal line to the right, twice as long. Complete the oblong, and draw its diameters. The height of the roof is equal to one-half of the height of the oblong. It should project 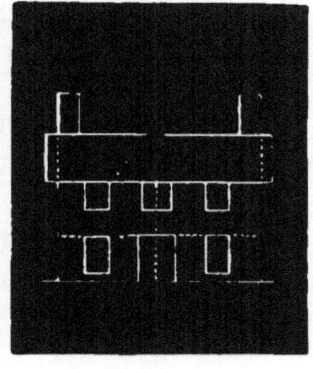 slightly at each end. Notice that the horizontal diameter of the oblong gives the tops of the door and of the lower windows. Add the windows and chimneys.

Card-Exercise VI.

Concentric Squares.

Draw a square on its diameters. Using the diameters of the first square for the diagonals of a second square, draw the second square inside the first. Divide the semi-diameters of the first square into two parts, and, through the points of division, draw a third square. Erase the dotted lines.

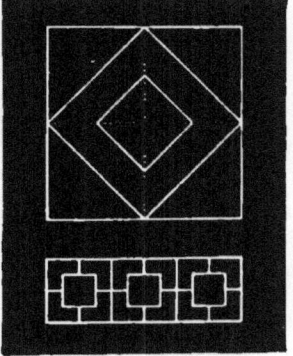

CONCENTRIC SQUARES REPEATED HORIZONTALLY.
— Draw two parallel horizontal lines of any designated length, and divide the intervening space into squares. Draw the diameters of the squares, and then divide each semi-diameter into two parts. Through the points of division on the semi-diameters, draw other squares. Erase those parts of the diameters of the larger squares which fall within the smaller squares.

These right-line figures are but applications of previous lessons. If any of their features are not understood by the pupils, refer them to those exercises where they were first drawn and explained.

CARD-EXERCISE VII.

A Four-pointed Star formed by a Square and Isosceles Triangles.

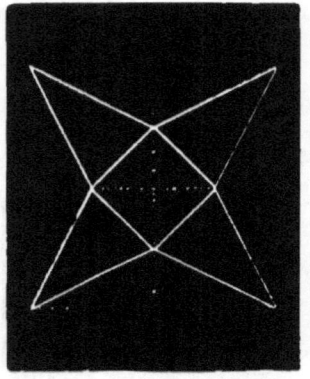

Draw a square of any definite size, and then add its diameters. Halve the semi-diameters, that is divide them into two equal parts, and on the points of division construct a square. On each side of the last square, draw an isosceles triangle, with its apex in a corner of the first square. Erase the first square and its diameters, indicated by dotted lines in the copy, and there is left a four-pointed star.

Call the attention of the pupils to the symmetrical arrangement of the lines about the centre of the square. Show them that when they have drawn one line they must draw certain other lines having the same position with reference to the centre of the figure. It will be found specially serviceable in dictation exercises, thus to make use of this fundamental principle of design,—symmetrical arrangement about a centre.

Card-Exercise VIII.

A House and Porch.

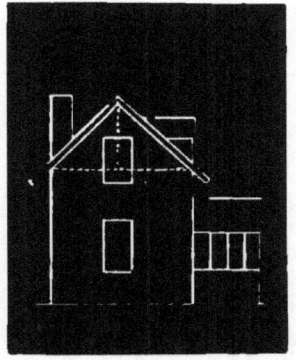

Draw a square for the body of the house. From the centre of the upper side of the square, draw a line upwards, equal to one-half of the side of the square. From the top of this line, draw the inner oblique lines of the roof, and then the outer parallel oblique lines. Notice that the roof projects beyond the sides of the house. Make the chimney as high as the ridge of the roof. On the other side of the roof, draw the horizontal and vertical lines forming the projection for the window. Dividing the side of the roof into four equal parts will give points where the lines of the gable strike the roof. The body of the porch is formed by a square drawn on one-half of the height of the side of the house. The roof of the porch is half as high as the body of the porch, and projects slightly beyond the body. Divide the upper half of the body of the porch into four parts. The lower window of the body of the house is half the height of the house, while the width of the window is half of its height. The dimensions of the upper window are a little less than those of the lower window.

It is not to be expected that young children will at first readily comprehend the various measurements of this exercise. It will be sufficient if they get an idea of the method of constructing a drawing by proportion. This exercise should be drawn as large as circumstances will permit.

Card-Exercise IX.

Erect Cross.

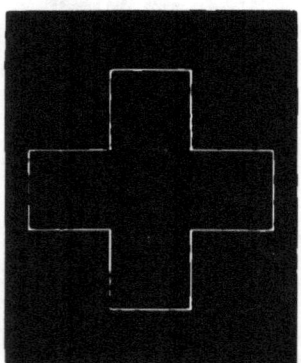

Draw a square, and divide each of its sides into thirds. Connect the points of division by two vertical and two horizontal lines, thus forming nine squares. Erase the central square and each corner square: in other words, erase the lines which are dotted in the copy.

Those lines which are drawn to help in the construction of the figures, and so are sometimes called "construction lines," are always to be erased when the drawings are finished. They are the dotted lines of the copies. Let it be remembered, then, that these lines are always to be erased.

This exercise, with some slight additions, such as the diagonals of the small squares, will answer well for a dictation exercise.

Card-Exercise X.

A Cross formed of Five Squares.

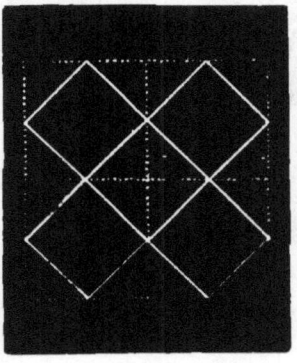

Draw a square on its diameters. Divide each of its sides into four equal parts. Divide the semi-diameters into two equal parts. Connect the points of division as in the copy, aiming to make the intersecting lines of the cross meet at the points of division on the diameters. In other words, connect the points of division by oblique lines, so as to form five oblique squares inside the first square.

Card-Exercise XI.

Two Oblongs in a Square.

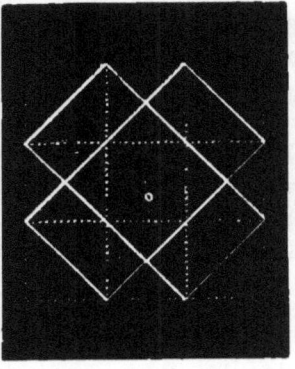

Draw a square, and divide its sides into thirds. Through the points of division, draw two vertical and two horizontal lines, making nine squares. In each corner square, draw a diagonal, with its side towards the centre of the large square. Lastly, draw lines connecting the ends of the diagonals, which are parallel. The teacher can easily modify this exercise, and make a dictation lesson, by adding other lines parallel or oblique to those already drawn.

Card-Exercise XII.

A Cross with its Angles all Acute.

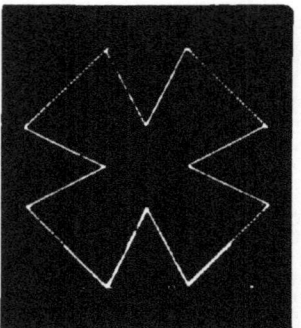

Draw a square, and divide each of its sides into thirds. Through the points of division, draw horizontal and vertical lines, forming nine smaller squares. Divide each side of the inner square into halves. In each corner square, draw a diagonal, with its side to the centre of the large square. From each end of each diagonal, draw a line to the nearest point of division on the inner square.

Card-Exercise XIII.

Right-Line Mouldings.

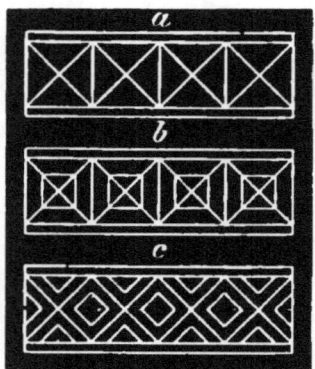

Form a. — Draw two parallel horizontal lines, at any given distance apart. Divide the space between them into squares, and draw the diagonals of the squares. Add a horizontal line above, and another below.

Form b. — Draw two horizontal lines, at any given distance apart. Divide the intervening space into squares, and draw their diagonals. Divide each half or semi-diagonal into two equal parts, and, through the points of division, draw the inner squares. Add a horizontal line above, and one below.

STRAIGHT-LINE FIGURES. 53

FORM *c*. — Draw two horizontal lines, at any definite distance apart, and divide the space between them into squares. Draw their diameters and diagonals. Divide each semi-diameter into two equal parts; and, from the points of division, draw lines to the sides of the squares. — these lines to be parallel with the diagonals. Add a horizontal line above, and one below.

A PAGE OF RIGHT-LINE FIGURES.

It is intended that the drawing of the common objects given on the following page shall be interspersed with the other exercises. These forms are to be drawn without directions, the teacher putting them on the board, and the pupils reproducing them on their slates. Observe that round as well as rectangular objects are represented by the straight lines. Similar outline drawings of other objects may be readily added to these. When doing these exercises the principal lines should be drawn first.

Thus far we have worked with figures composed of straight lines alone. By this time the names and character of straight lines, and of various straight-line figures, should be firmly fixed in the minds of the pupils. The next exercises will include curved lines, the drawing of which will give more varied and interesting practice.

QUESTIONS. — What is said of reduction and enlargement in drawing? Of class-analysis of forms before beginning to draw them? Describe memory drawing. What is its use? When are squares said to be concentric? What is the use of the dotted lines in the copies? and what is to be done with them?

54 TEACHERS' MANUAL.

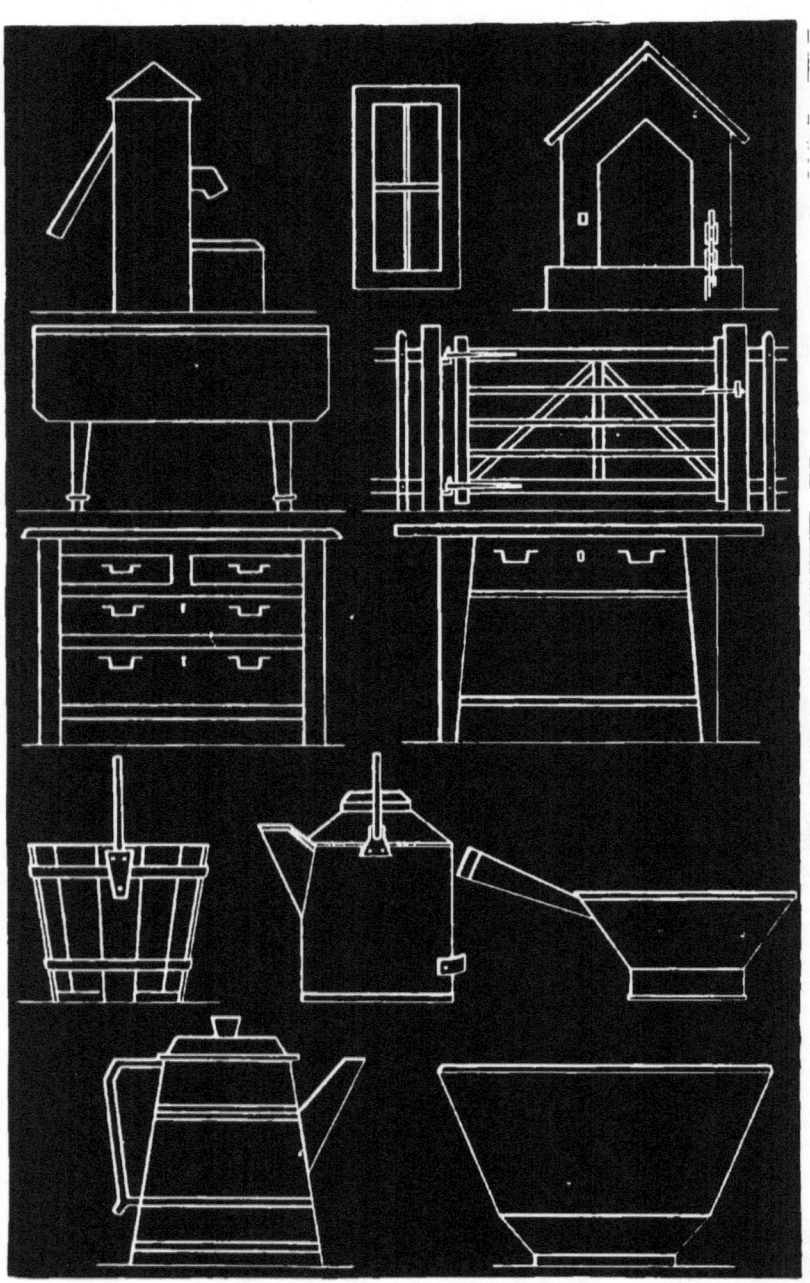

CHAPTER III.

SIMPLE CURVES.

IF the pupils have been interested in the preceding exercises, they will be much more interested in the ones which follow. It is rare to find children that do not like to draw. As soon as they acquire a little skill in the use of their pencils, they delight to apply it in every direction. While they should be allowed great freedom in the exercise of their skill, the teacher should frequently require them to submit their independent work for inspection, in order that the faults may be pointed out and corrected.

A Curve.

A curved line is a line which changes its direction at every point.

Curved lines are of different kinds.

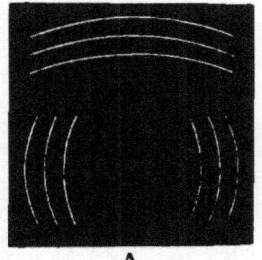

A.

A SIMPLE CURVE.— *A simple curve bends regularly, and, if continued, will form a circle.*

All simple curves, less than a circle, are but parts of a circle. The curves at A are simple curves.

The teacher should have suitable objects for the

purpose of illustrating the different kinds of curves. In the set of models for primary schools, there is a disk with which the various features of the circle, or parts of circles, may be illustrated.

Having drawn several simple curves on the board, and explained them to the pupils, require the latter to draw from memory a right angle, and then to connect the two ends of the lines forming the angle by a simple curve, as at B.

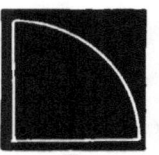

This figure is called a quadrant, because it is one-quarter of a circle. Repeating the curve to the left, we have a semi-circle (C), or one-half of a circle. Repeating the lines below, we have the complete circle (D), with

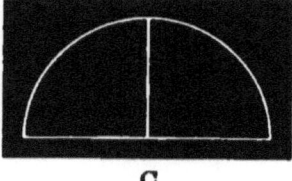

its horizontal and vertical diameters.

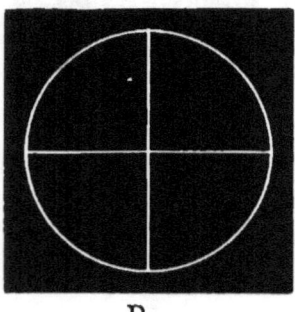

To draw a semi-circle (C), begin by drawing a horizontal line; at the centre of this, draw a perpendicular, equal to one-half of the horizontal line. Connect the ends of these lines by a simple curve. To draw a circle (D), begin by drawing two of its diameters, the same as for a square; then connect the ends of these lines by four simple curves, which will form, when all drawn, one continuous simple curve, or circle.

It is not to be expected, at first, that pupils will, in this way, draw a circle with any great degree of accuracy. Indeed, it is not here given so much as an exercise to be copied by the pupils, as for the purpose of showing them the completion of a simple curve, — showing that all simple curves, if continued, will form circles.

When a curve is drawn on a straight line, as at E, the straight line is called the base of the curve, while the distance of the curve from its base is called the altitude, or height, of the curve. The dotted line in E shows the altitude of the curve at its centre.

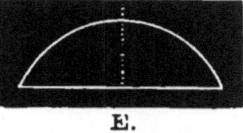
E.

It is important that pupils have a clear idea of the base and altitude of a curve. The base is a straight line connecting the ends of a curve; it may be vertical, horizontal, or oblique, as shown at F.

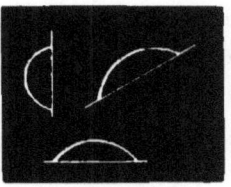
F.

We do not say a vertical curve, or a horizontal curve, or an oblique curve; but we describe curves by their base lines, as "a simple curve on a vertical base," or, "a simple curve on a horizontal base," &c. The altitude of a curve, or of any point of a curve, is its perpendicular distance from the base. It will be remembered that perpendicular does not necessarily mean vertical. Hence every curve has altitude, whether its base line be horizontal, vertical, or oblique. If a curve has a vertical base, the distance of the curve from its base is called the altitude or height of the curve, just as if it had a horizontal base. When we simply speak of the altitude of a curve, we mean its greatest altitude;

but we may speak of its altitude at any point, besides the highest, provided we designate the point.

Draw on the blackboard curves with vertical, horizontal, and oblique bases, and with their centres at different altitudes, until pupils have a clear comprehension of the terms; for these terms will be frequently used in succeeding exercises.

In the page of skeleton letters will be found some good forms for blackboard copies. As the letters afford the simplest combinations of straight lines and curves, and as the forms are familiar, no definite proportions need be given here: they will be readily discovered. If the subsequent exercises should be interspersed with the drawing of these letters, it will give a pleasing variety to the work. Not too much of the same thing at one time.

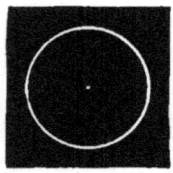

A CIRCLE. — *A circle is a plane figure bounded by a curved line, called its circumference, every part of which is equally distant from a point within, called its centre* (G).

This is the strict mathematical definition of a circle, and makes it consist of the space enclosed by the curved line.

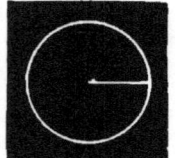

A RADIUS. — *A radius of a circle is a straight line drawn from its centre to the circumference* (H).

When we wish to speak of more than one radius, we say *radii*. A circle may have an unlimited number of radii; that is, you may

draw any number of radii, and yet there can be one more drawn afterwards.

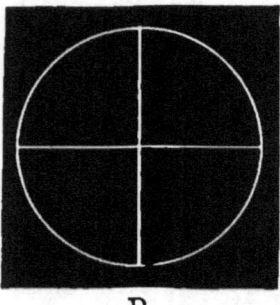

D.

DIAMETER OF A CIRCLE. — *A diameter of a circle is a straight line, drawn through the centre and touching the circumference on both sides* (D).

A diameter of a circle is made up of two radii running in opposite directions. There may be an unlimited number of diameters, as there may be an unlimited number of radii. Radius means spoke; diameter, that which measures through.

These, then, are the features of the circle: *The circumference*, which bends equally in all its parts; *the centre*, which is at the same distance from every point in the circumference; *the radius*, or distance from the centre to the circumference; *the diameter*, which is the longest straight line that can be drawn in the circle, and divides the circle into halves, called semi-circles.

In popular usage, the word "circle" is often employed simply to indicate the curved line bounding the circle, and might be defined thus: —

A circle is an endless curved line, which bends equally in every part.

CHORD AND ARC OF A CIRCLE. — A straight line which touches the circumference of a circle at two points, but is shorter than the diameter, is called a chord. It divides the circle into two unequal parts, either of which is called an arc. In drawing, the chord is usually called the base of the curve.

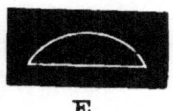

E.

Whenever children are given definitions, the definitions should be, if possible, strictly accurate, that there may be nothing to unlearn afterwards. The scope of the definitions should be repeatedly pointed out and illustrated. In the present instance, the mathematical definition of a circle might be illustrated by a penny, while the popular usage of the word might be illustrated by a ring. In the one case, the whole of the penny would represent the circle; in the other case, the circle would be represented by the ring. Thus the distinction between the two usages of the word would be clearly seen.

The pupils should be required to name objects which are circular in form. They would doubtless mention wheels, hoops, plates, &c.; quite likely they would also mention balls, oranges, pencils, &c., because these objects have the element of roundness. At this point, therefore, it would be well to give some special attention to the word "round." It is a long time before children learn, without some special instruction, the exact force of this common word. Various objects they can only describe as round, though their forms may differ in a marked degree. Thus, a circle like a ring, a circle like a disk, a sphere, a cylinder, a cone, are all round, and nothing more. But with a few illustrations, repeated several times, they soon comprehend that the word "round" is insufficient to describe all objects whose forms have the element of roundness; and, when the distinctions between the objects are once clearly seen, they readily learn to use the words which exactly express these distinctions.

CARD-EXERCISE XIV.

Inscribed and Circumscribed Circle. — Globe. — Wheel.

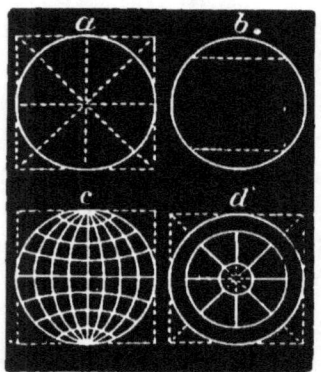

FORM *a.* *A Circle inscribed in a Square.* — In a previous exercise (D) we drew the circle on its horizontal and vertical diameters; we will now draw a circle in a square. Having drawn a square, with its diameters and diagonals, mark on the diagonals points through which the circle should pass. Connect these points and the ends of the diameters by simple curves. Draw the curves carefully, section by section. A circle drawn inside a square, and touching the square, is said to be inscribed. Erase the dotted lines.

FORM *b.* *A Circle circumscribed about a Square.* — Draw a square; then, on each side of the square, draw a simple curve. When the simple curves are united at the corners of the square, they should form one endless simple curve or circle. When a circle is thus drawn around a square, it is said to be circumscribed.

FORM *c.* *A Globe.* — Draw a square and its diameters; then an inscribed circle. Divide each semi-diameter and each quarter of the circle into four

SIMPLE CURVES. 63

equal parts. Through the points of division, draw the inner curves.

FORM *d*. *A Wheel.* — Draw a square, with its diameters and diagonals; and then an inscribed circle. Divide the semi-diameters of the circle into four equal parts; through the outer points of division, draw a second circle, and another through the inner points of division. Erase the dotted lines.

CARD-EXERCISE XV.

A Cross with Angles, Right, Acute, and Obtuse. — Fret Moulding.

Draw a square, and divide it into nine equal smaller squares. Divide each side of the inner square into two equal parts. From each point of division on the inner square, draw oblique lines to the points of division on the corresponding side of the large square.

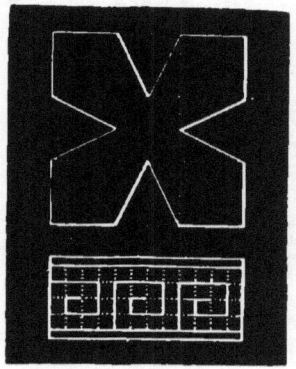

FRET, OR KEY, MOULDING. — Draw three squares, as in Card-Exercise III., united horizontally. Divide the left side of the left square into four equal parts, and, from the points of division, draw horizontal lines dividing all the squares. Divide the upper side of each square into four equal parts, and, from these points of division, draw vertical lines through the squares, dividing each of the first squares into sixteen smaller squares. Draw the lines forming the fret as shown. If the pupils are, unfortunately, without cards, then the teacher must give further verbal explanation, or illustrate on the board. Add a horizontal line above, and another below. This is one of the most popular, as it is one of the oldest, ornamental forms ever invented.

Card-Exercise XVI.

Rosettes composed of Simple Curves.

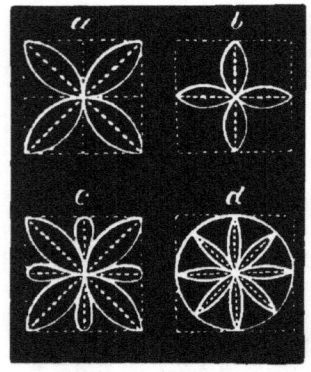

Form *a.* — Draw a square on its diameters, and add the diagonals. Using each semi-diagonal as a base, draw a simple curve on each side. Aim to make all of the curves alike.

Form *b.* — Draw a square on its diameters. Using each semi-diameter as a base, draw a simple curve on each side.

Form *c.* — Draw a square, with its diameters and diagonals. Using each semi-diagonal as a base, draw a simple curve each side. Add the curves on the semi-diameters.

Form *d.* — Draw a square, with its diameters and diagonals. Within the square, draw an inscribed circle; using each semi-diameter of the circle as a base, draw a simple curve on each side.

Card-Exercise XVII.

An Eight-pointed Star and Fret Moulding.

Draw a square, and divide it into nine equal small squares. Divide each side of the central square into

SIMPLE CURVES. 65

two equal parts. From the point of division on each side of the small square, draw oblique lines to each point of division on the corresponding side of the large square. Using the diameters of the central square as diagonals, draw an inner square. Remember that the lines which are dotted in the copy are always to be erased.

FRET MOULDING. — Draw a series of squares, as for the moulding in Card-Exercise XV. Divide the left side of the left square into five equal parts, and, from the points of division, draw horizontal lines through all of the squares. Divide the upper side of each square into five equal parts, and, from the points of division, draw vertical lines dividing each of the large squares into twenty-five small squares. Draw the fret as in the copy, and add the horizontal lines.

CARD-EXERCISE XVIII.

Quatrefoil and Trefoil. — Waterbottle and Moulding.

The syllable *foil*, frequently used in design, is from the French, and means leaf, as the teacher should explain to the children; for that will help to fasten the term in their minds. Trefoils (three leaves), quatrefoils (four leaves), cinquefoils (five leaves), are forms common in nearly all styles of modern ornament.

FORM *a.*   *Quatrefoil on a Square.* — Having 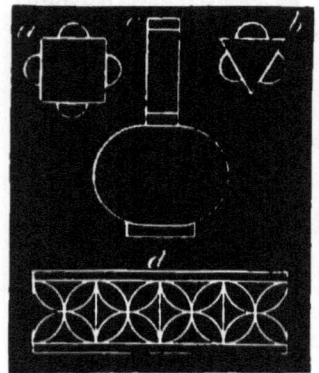 drawn a square, divide each of its sides into four equal parts. On the two inner parts, as a base, draw a simple curve, with an altitude equal to one-half of the base. These curves are foils; and, as there are four of them in this instance, we have what is called "quatrefoil on a square."

FORM *b.*   *Trefoil on a Triangle.* — Draw an equilateral triangle. Divide each side into four equal parts. On the two inner parts of each side, as a base, draw a simple curve or foil, with an altitude equal to one-half of the base.

FORM *c.*   *A Water-Bottle.* — Draw the central vertical line, and divide it into six equal parts. Add the two other vertical lines, one-sixth of their length apart. On the lower halves of the outer lines as a base, draw simple curves, with an altitude equal to one-half of their base. Connect the ends of these curves by horizontal lines; add the bottom and the short horizontal lines above.

FORM *d.*   *A Moulding composed of Simple Curves.* — Draw two parallel horizontal lines, of any length, and of any given distance apart. Divide the intervening space into squares, and draw their diagonals. Where the diagonals cross each other will, of course, be the centre of the square. On each side of each semi-diagonal, draw a simple curve. Add a horizontal line above, and another below.

SIMPLE CURVES.

This moulding, whose unit is the same as A in Card-Exercise XVI., illustrates two principles of design, — symmetrical arrangement about a centre, and horizontal repetition of the unit.

Card-Exercise XIX.

Lozenge Moulding. — Simple Curves on a Square, and Simple Curves in a Square.

Form *a*. — Draw three or more squares, united horizontally. Draw their diameters, whose points of intersection will give the centres of the squares. On each upper and lower side of each square, as a base, draw a semi-circle through the centre. Add a horizontal line above, and another below.

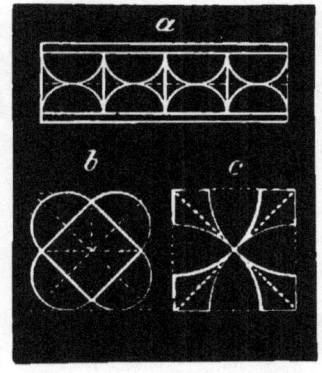

Form *b*. *Semi-circles, or Simple Curves, on a Square.* — Draw a square on its diameters. Using the larger part of these diameters as diagonals, draw an inner square. On each side of the inner square, draw a semi-circle. Erase all but the inner square and the semicircles.

Form *c*. *Simple Curves in a Square.* — Draw a square and its diagonals. Divide each side of the square into three equal parts. Connect the upper division of the left side of the square with the right division of the lower side of a square by a simple curve drawn through the centre of the square. Draw the three corresponding curves.

Card-Exercise XX.

Four-pointed Star, Outline.

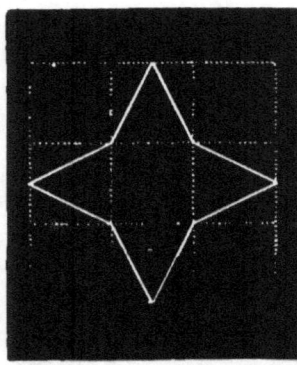

Draw a square, and divide it into nine squares. On each side of the inner square, draw an isosceles triangle, with its apex in the centre of the corresponding side of the large square. Erase all except the side-lines of the triangles.

Card-Exercise XXI.

A Bird-House.

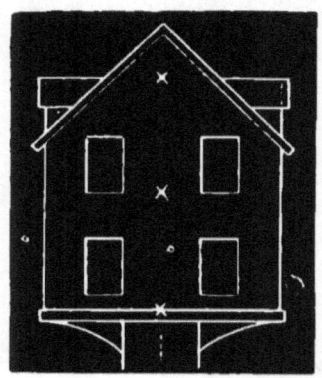

Draw a square and its diameters. On this, as a base, construct the house, without further directions. This is a good figure for class analysis.

Double Squares and Crosses. — Repeating Form for covering Surfaces.

Draw three or more parallel horizontal lines at the same distance apart. Draw three or more vertical lines, crossing the horizontal lines, and forming a series of squares. The whole slate or blackboard may thus be covered with squares. Divide each side of each square into two equal parts. Draw oblique lines, connecting the points of division and forming a square within each

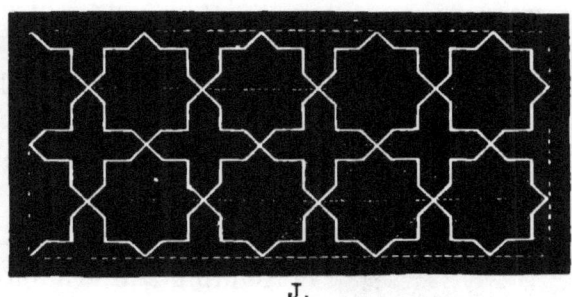

J.

of the original squares. Divide each side of the last formed squares into three equal parts, and, through the points of division, draw lines parallel with the sides of the original squares, and forming a third series of squares. Erase all the lines of the original squares, and the central third of each side of the last squares. There will be left a regular series of double squares and crosses.

The same result may be reached in this way, avoiding the first series of squares. Draw a square, or an oblong, of definite size; an oblong, for example, four inches by six. Divide the sides into equal parts, an inch or two inches long, for instance. Beginning with the points of division nearest the corners, connect all the points by oblique lines, forming a series of oblique squares. Finish as in the first mode.

Direct the attention of pupils to similar forms to be found in carpets, wall papers, &c.

Semi-circles and Straight Lines for covering Surfaces.

K.

Draw a series of squares to fill the slate or blackboard. Divide each side of each square into four equal parts. On the two central parts of each side, draw a semi-circle, tending alternately towards and from the centre of the square.

CARD-EXERCISE XXII.

Forms for Repetition, composed of Simple Curves.

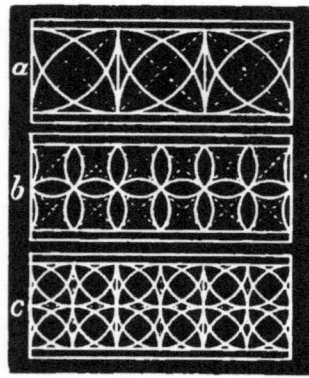

FORM *a*. — Draw two parallel horizontal lines of any length, and at any given distance apart. Divide the space between them into squares. Draw the diagonals of the squares, and divide each semi-diagonal into two equal parts. Connect the opposite ends of each diagonal by simple curves drawn through the divisions of the other diagonal of the same square. Add a horizontal line above, and another below.

FORM *b*. — Draw a series of squares, united horizontally, as in A. Draw their diameters and diagonals. Connect the ends of the horizontal diameters to the ends of the vertical diameters by semi-circles drawn through the centres of the squares. Taking each half of the vertical lines which separate the squares as a base,

SIMPLE CURVES. 71

draw on each side of the base simple curves, to unite with the semi-circles. Add a horizontal line above, and another below.

FORM *c.* — Draw a series of squares, as in the last two exercises. Draw their diameters. On each side of each square, as a base, draw a semi-circle through the centre of the square. Then draw a circle, passing through the ends of the diameters.

As an addition to these exercises, draw a similar series of squares above, and another below, each united with the first series. Fill these squares, in the same manner as the first were filled, with curves. If they are correctly drawn, they will in each case form an involved series of perfect circles. Omit the parallel horizontal lines.

These exercises for covering a surface will be found good ones for the pupils to practise frequently. Much of what has been learned in previous lessons will be brought into use. The pupils should be permitted to cover their slates with such forms, and also to draw them with many repetitions on the blackboard. If a slate is filled with them carefully, there ought to be considerable improvement between the first and the last lines drawn.

QUESTIONS. — Describe a curve. A simple curve. A quadrant. A semi-circle. What are the features of a circle? What is the base of a curve? The altitude? What is a chord? An arc? What is the difference between the mathematical and the popular use of the word "circle"? How should definitions be given to children? What of the word "round"? When is a circle said to be circumscribed about a square? Inscribed in a square? What is meant by "foil"? Trefoil? Quatrefoil?

CHAPTER IV.

COMPOUND CURVES,—THE ELLIPSE AND OVAL.

Thus far we have dealt with simple curves; that is, with curves struck from one centre, and having, consequently, the same degree of curvature in every part. All simple curves are parts of circles, and, if continued far enough, become circles. A circle, therefore, is but an endless simple curve. A compound curve is a curve that does not bend regularly throughout its whole length, and must, therefore, be struck from two or more centres. As the curve is drawn, the degree of curvature may change at every point, as in the ellipse; or it may change at intervals, as in the oval; while it may completely reverse its direction, as in the ogee.

The Ellipse.

The ellipse is a plane figure bounded by a compound curve struck from two centres (A).

There are many curious things to be said about the ellipse. It is the form a circle appears to take when viewed obliquely; and a circle is always viewed thus, except when the eye is directly opposite its centre. Draw a circle on the blackboard: step to one side and view it, and the circle will apparently become an ellipse. Draw a circle on a slate; turn the slate from the eye, and view the circle: it

COMPOUND CURVES. 73

will then take the form of an ellipse. Hold a hoop in a vertical position, with its centre directly in front of the eye: the hoop will then be seen as a circle, — its real form. Turn the hoop from the eye, and its form will appear to have changed to that of an ellipse. The farther the hoop is turned from the eye, the flatter will become the ellipse, until its two sides merge in one, and only a straight line is seen. Take a round stick: cut it across, obliquely and evenly; the shape of the section or scarf will be that of an ellipse. The more obliquely the stick is cut, the greater will be the length of the section compared with its breadth; but it will always be an ellipse.

A disk or cone affords the best means for illustrating the ellipse; but, if the proper models are wanting, the teacher can, by using the modes of illustration just described, familiarize the pupils with the exact form of the ellipse. It is essential that they should become perfectly familiar with this curve, as it is so frequently employed by both artists and artisans.

To draw a cone, with an oblique view of its base, first draw an ellipse, and then add a vertical and two oblique lines, as shown at B. The oblique lines should not be drawn to the ends of the long diameter, but tangent to the curve. Be particular to observe this. Erase the diameters. Draw several

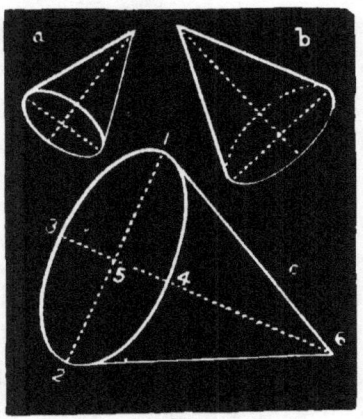

B.

cones, giving a more or less oblique view of the base. Illustrate with a model.

There are many variations in the form of the ellipse, since its form depends on the difference between its length and its breadth. It may be so broad as closely to approach a circle, or so narrow as to approach closely a straight line; but, whatever its form, it always retains certain distinctive features.

Both its long, or *transverse* diameter, and its short, or *conjugate* diameter, divide the ellipse into two similar equal parts. The two diameters, bisecting each other at right angles, divide the ellipse into four similar equal parts.

If it is required to find the two centres of an ellipse already drawn, then draw from an end of the short diameter two oblique lines, one on either side, to the long diameter; each of these oblique lines having a length equal to one-half of the long diameter. Where the oblique lines touch the long diameter will be the two centres, or *foci*, of the ellipse. Now take any point in the circumference of the ellipse, and draw a line from that point to each centre of the ellipse; and these two lines united will equal the length of the long diameter. If that diameter is two feet, for example, the two lines will measure two feet. If these two oblique lines do not equal the length of the long diameter, then it follows that the ellipse is not correctly drawn.

Large ellipses, of different proportions, should be drawn on the blackboard, from time to time, for the pupils to study. This can be easily done. Take a piece of thread; tie a loop at each end; stick two pins into the blackboard, at a less distance apart than the length of the thread; put a loop over each pin, when the thread will hang loosely; press a crayon against the thread until it is rendered taut; then carry the crayon around, touching the blackboard, and making an ellipse.

If an ellipse is required having diameters of definite length, as a long diameter of three feet and a short diameter of two feet, draw the diameters, and find the centres of the ellipse, as already described. Placing an end of the thread at each of these centres, with slack enough to allow the crayon, when pressed against the thread, to touch an end of the short diameter, draw the ellipse, when its boundary-line will pass through the ends of both diameters, and be of the required proportions.

By drawing the ellipse in this manner, the children will readily perceive what is meant when it is said, that the ellipse is a compound curve, struck from two centres. As the two pins, or two centres, are brought nearer and nearer together, they will readily perceive that the ellipse approaches nearer and nearer a circle, while it never becomes a circle until the two loops are placed over one pin and the curve is struck from one centre. Literally no figure can have two centres.

Children need not be taught all that is here said about the ellipse before they begin to draw it. Indeed, it is best for them to acquire a knowledge of its features along with their drawing, and so acquire that knowledge gradually. They should, however, have a distinct idea of the form of an ellipse before they begin to draw one freehand. This they can obtain by viewing a circle obliquely, by viewing an actual ellipse, either drawn on the board or shown by a model. In either of the last two cases, the eye must be directly opposite the true centre, or point where the two diameters intersect.

To draw an ellipse freehand, as at A, begin by drawing a vertical line for the transverse, or long diameter. Divide this line into two equal parts; through the point of division, draw a horizontal line for the short diameter, having one-half the length of the long diameter. Through the ends of these diameters, draw the boundary-line of the ellipse. See that no part of this boundary-line is straight; that it nowhere makes a sharp turn; that each quarter of the ellipse has the same shape as the other three-quarters. Require the pupils to draw the same on their slates, making the long diameter four inches, for example, and the short diameter two inches.

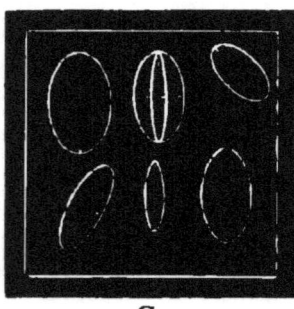

C.

After several ellipses, with the proportions 1 × 2 (one by two), like the one at A, have been drawn, then draw others, with different proportions, and in different positions, as at C. Require the pupils to reproduce them on their slates.

The path of the earth about the sun is an ellipse. The sun is in one of its foci; but, were an ellipse drawn on the blackboard having the same proportions as the orbit of the earth, it would be, perhaps, impossible for the unaided eye to tell it from a true circle.

AN ARTIST'S PALETTE. — Draw an ellipse; add the handle and thumb-hole. The marks on the palette show where the color is usually placed.

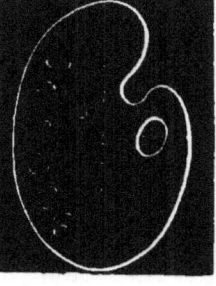

A celebrated painter was once asked with what he mixed his colors. "With brains, sir," he replied. Indeed, without brains, without thought, nothing can be well done.

The Oval.

The oval is a plane figure bounded by a compound curve only one of whose diameters crosses the other at the centre. — Draw a vertical line; divide it into three equal parts; through the upper point of division, draw a horizontal line, as at D, equal to two-thirds of the vertical line. One-

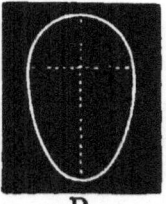

D.

half of the horizontal line extends on either side of the vertical line. On the horizontal line, as a base, draw a semi-circle through the upper end of the vertical line. On the lower two-thirds of the vertical line, draw one-half of an ellipse, to unite with the ends of the semi-circle. This gives an oval, or egg-shaped figure, as the word means, one part of whose boundary-line is a simple curve, the other part an ellipse.

It will be seen that the long diameter divides the oval into two equal similar parts, and that the short diameter does not. We say two *equal similar* parts. "Equal" refers to the area, or surface; "similar" refers to the shape. Hence two equal figures may have different shapes; while two similar figures may have different areas. As the two parts of the oval, when divided by its long diameter, should be alike in both respects, it is necessary to describe them as equal similar parts. Either diameter of an ellipse, which is sometimes called an oval, but is not an oval, divides it into two equal similar parts.

It is not essential to an oval that its boundary-line be in part a circle, and an ellipse in part, like D. On the other hand, it may be composed wholly of parts of different circles, or wholly of parts of different ellipses ; or it may have no part of a circle or an ellipse, but be composed wholly of other curves. It is only essential that it retain the shape of an egg ; that the long diameter divide it into two equal similar parts, while the short diameter does not. Draw on the blackboard various differently proportioned ovals, and require the pupils to draw them on their slates. Do not expect that they will at first draw them with any great degree of accuracy: be satisfied if they get the general idea.

By this time the pupils should have a clear comprehension of the circle, the ellipse, and the oval, those three important curve-figures. They should be able to describe the features of each separately, and to contrast one with another. Do not, however, drop the figures here : revert to them frequently by way of review. When new figures containing curves are to be drawn, require the pupils to examine them carefully, to see what curves enter into their composition. Class-analysis of figures adapted to the purpose should frequently include a general review of these curves.

COMPOUND CURVES.

Mallet.

Draw a vertical line, and divide it into two equal parts. On the upper half draw the head of the mallet; on the lower half, the handle. The greatest width of the head is one half greater than its height. It will be seen that the curves forming the sides of the head are compound, like parts of an oval.

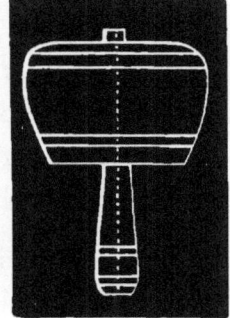

Card-Exercise XXIII.

Hour-Glass, Kite, Top.

An Hour-Glass. — Draw the central vertical line, and divide it into three equal parts. Draw the horizontal lines, and then the side lines, making the width of the glass equal to one-third of its height. The thickness of the frame is equal to one-sixth of the width of the glass; while the top and bottom project from the glass one-third

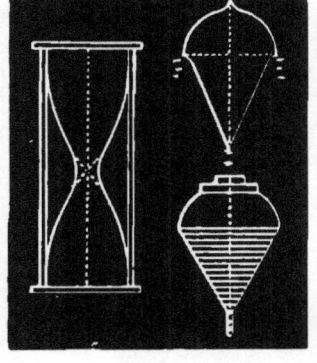

of its width within the oblong frame. Draw the compound curves, extending two-thirds of the length of the frame, and making the waist of the glass equal to one-third of its greatest width.

A Kite. — Draw the central vertical line of any given length, and divide it into three equal parts.

Through the upper point of division, draw a horizontal line two-thirds as long as the vertical line. On this horizontal line, as a base, draw a semi-circle through the top of the vertical line. From points near the ends of the horizontal line, draw oblique lines to the bottom of the vertical line. Add the point at the top, and the slight curves at the upper ends of the oblique lines.

A Top. — Having drawn a vertical line of any given length, divide it into three equal parts. Through the upper point of division, draw a horizontal line equal to two-thirds of the vertical line. Divide the lower third of the vertical line into halves, and, through the point of division, draw a very short horizontal line. Connect the ends of the horizontal lines, on either side, by an oblique line, forming the straight sides of the top. Draw the horizontal lines at the top, making the upper one somewhat less than a third of the vertical line. Add the curves, the remaining horizontal lines, and the peg at the bottom.

Acorn and Cup. — Ovoid Form.

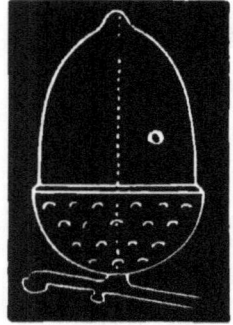

Draw a vertical line, and divide it into three equal parts. Through the lower division, draw the longest horizontal line, equal to two-thirds of the vertical line. Upon this skeleton construct the figure. The cup is a semi-circle; the acorn a semi-ellipse; the whole is an oval. Add the stalk at the bottom.

Card-Exercise XXIV.

Lemon and Barrel. — The Ellipse Illustrated.

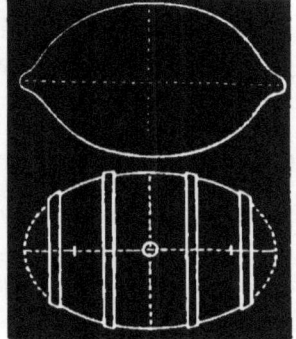

LEMON. — First draw an ellipse, with its short diameter about two-thirds of the length of the long diameter. Add the points forming the ends of the lemon.

BARREL. — This is a good figure for class-analysis before drawing. Thus, for example: —

Teacher. — What lines are seen in this figure?
Pupil. — Straight lines and curves.
Teacher. — What is the position of the straight lines?
Pupil. — Vertical.
Teacher. — If vertical, what else must they be, as they are side by side?
Pupil. — Parallel.
Teacher. — What do they represent in this drawing?
Pupil. — Hoops on a barrel.
Teacher. — But hoops are circular: these lines are straight. How is that?
Pupil. — When a hoop is turned so we cannot see through it, then it looks like a straight line.
Teacher. — Is that the position of the hoops on this barrel?
Pupil. — Yes.
Teacher. — If you hold a hoop with its centre opposite the eye, with each point of the hoop at the same distance from the eye, what will its shape appear to be?
Pupil. — A circle; and that is its real shape.
Teacher. — What is a circle?

Pupil. — A simple curve. It is struck from one centre, and bends alike in all parts.

Teacher. — When a hoop is viewed obliquely, what does its shape appear to be?

Pupil. — An ellipse.

Teacher. — What is an ellipse?

Pupil. — A compound curve struck from two foci.

Teacher. — Are the curves in this drawing simple, or compound?

Pupil. — Compound; because the sides of the barrel are parts of an ellipse.

Teacher. — How does the transverse, or long diameter divide the ellipse?

Pupil. — Into two equal similar parts.

Teacher. — Why do you say *equal* and *similar*?

Pupil. — Because "equal" refers to area only; "similar" to shape only. Two figures may be equal in area, but of different shapes; or they may be similar in shape, but of different areas.

Teacher. — How does the conjugate, or short diameter divide the ellipse?

Pupil. — Into two equal similar parts.

Teacher. — How do both diameters divide the ellipse?

Pupil. — Into four equal similar parts.

Teacher. — How do the diameters divide each other?

Pupil. — Into halves.

Teacher. — How do you find the two foci, or centres, of the ellipse?

Pupil. — By drawing an oblique line from an end of the short diameter, on either side, to the long diameter, the oblique line being equal to one-half of the long diameter. The points where the oblique lines touch the long diameter are the foci of the ellipse.

Teacher. — On what does the shape of the ellipse depend?

Pupil. — On the difference in the length of its diameters.

COMPOUND CURVES. 83

Teacher. — The less the difference? —
Pupil. — The more nearly the ellipse comes to being a circle.
Teacher. — The greater the difference? —
Pupil. — The more nearly it comes to being a straight line.
Teacher. — How many shapes may an ellipse have?
Pupil. — An endless number.
Teacher. — But what may be said of any point in the boundary-line of any ellipse?
Pupil. — That, if a line be drawn from this point to each of the foci, or centres, the two lines united will be equal to the long diameter.
Teacher. — In drawing an ellipse freehand, what is to be observed?
Pupil. — Not to make any part straight, nor any part a circle, nor a sharp turn anywhere.
Teacher. — Is it ever proper to call an ellipse an oval?
Pupil. — No; for an oval is always egg-shaped: one end is always larger than the other, which is not true of an ellipse. The short diameter of the oval never divides it into two equal similar parts; but the short diameter always divides the ellipse in that way.

It is not expected that the teacher will follow exactly this mode of analysis. The manner of putting the questions, and their number, must be adapted to the capacity of the pupils.

After the preliminary analysis, bringing out many or few of the characteristics of the ellipse, proceed to draw the barrel thus: —

Draw an ellipse, with its short diameter about two-thirds of the long diameter. On either side of the centre of the long diameter, mark off a distance equal to one-half of the short diameter; divide what remains into halves, and, through the points of division, draw the heads of the barrel. Draw the end-hoops, making their width equal to one-half of the space obtained by the last divisions. Divide the space between the end-hoops into four equal parts; draw the bung in the centre, and the other hoops on the other points of division.

Bottle. — Ovoid Form.

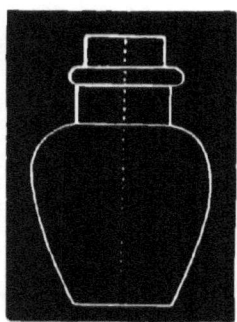

Draw a vertical line, and divide it into three equal parts. At the bottom, and through the upper point of division, draw horizontal lines, having a length equal to one-third of the vertical line. Draw the profile of the body, which is an oval slightly modified, and add the neck and cork.

Card-Exercise XXV.

Ellipse and Oval Illustrated. — Pear, Grapes, and Egg.

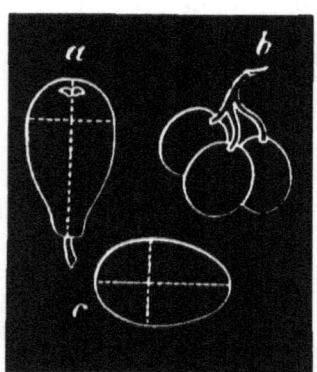

The pear (a), which is oval, the grapes (b), which are elliptical, and the egg (c), which is oval, are to be drawn by judgment of eye. The teacher will draw them thus on the blackboard, and will then require the pupils to draw them on their slates without definite proportions being given. Draw the full outline of each grape at first, and then erase the parts which should not show.

COMPOUND CURVES.

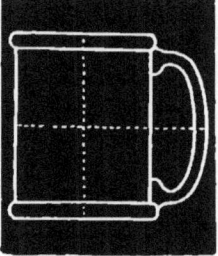

Mug. — The vertical and horizontal dotted lines are equal in length, while the handle is nearly one-third of the width of the horizontal line. It will be observed that the curves forming the handle have a semi-elliptical character. This is a favorite curve with Nature and with designers.

Card-Exercise XXVI.
Rosette of Ovoid Forms.

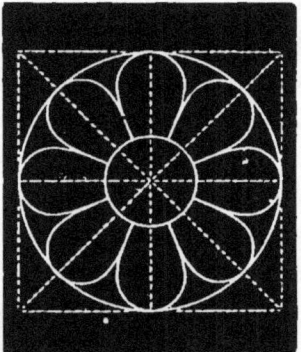

Draw a square, its diameters and diagonals; and then an inscribed circle. Divide each diameter into six equal parts, and, on the central divisions, draw the inner circle, which will thus have a diameter equal to one-third of the diameter of the large circle. On each part of the diameters and diagonals, between the two circles, draw an oval, making eight in all, and touching one another.

Questions. — What is a compound curve? An ellipse? What is the shape of a circle when viewed obliquely? What are the diameters of an ellipse called? How does each divide the ellipse? What is meant by the foci of an ellipse? How can they be found? How can you draw an ellipse of given proportions? What is an oval? How do its diameters divide it? Of what may its boundary line be composed? What is meant by "equal" and "similar" when applied to figures?

CHAPTER V.

COMPOUND CURVES.

REVERSED CURVES. — THE OGEE. — ABSTRACT CURVES.

THE compound curves to which we have attended are curves that change their direction irregularly ; but, when completed, they enclose space, and the names given them are sometimes used to indicate the figures thus formed, and sometimes to indicate the boundary-lines of the figures. In this way the word ellipse may be applied to the space enclosed, or it may be limited to the boundary-line enclosing the space.

We will now attend to those curves which not only change their direction irregularly, but reverse it, turning from the right to the left, from the left to the right, and never enclosing space. The two parts of this form of the compound curve may each be a simple curve struck from one centre, or a compound curve struck from two or more centres.

Draw two equal circles, as at A, touching each other. Through the point of contact, draw a straight line, dividing each circle into two unequal parts. Erase the two larger parts, and there will remain a compound reversed curve as at B, each half of which is a simple curve, being the arc of a circle. It will be seen that this compound curve has the same degree of curvature in all its parts, but that the direction of the curve is instantly reversed at the point where the straight line divides it.

Instead of parts of circles, the reversed curve may be composed of parts of ellipses or parts of ovals. Each of the two sections of the curve would then be compound in itself. Again, each segment of the reversed curve may be quite different from any one of the compound curves which have yet been described. It may be of such a character that it can be struck by mechanical means, or it may be what is sometimes termed a "hand-curve," because it has no mathematically-defined features, and is struck by hand without the aid of instruments. To this last variety belong the first three curves shown at C.

The first curve at C is called the ogee, and sometimes the "line of beauty." Its two segments are alike. The degree of curvature of each is greatest at the point of its greatest altitude, and grows less
and less from that point. As we approach the point where the curve changes its direction, it becomes nearly a straight line. Each segment, therefore, is a finely-graduated compound curve; and the two segments unite without any sudden change of direction in the curve. Let this be carefully noted and remembered. Where the segments of the compound curve at B unite, the direction of the curve is instantly and completely reversed. This

makes a somewhat disagreeable impression upon the eye; but, where the segments of the first compound curve at C unite, there is no such abrupt change of direction, and the impression made upon the eye is consequently agreeable. This is something to be observed in drawing compound curves freehand, if we would make them beautiful.

Let the ogee be repeatedly drawn, until pupils are familiar with its features. First, draw a straight line for the base, and halve it. Beginning at one end of the base, draw the curve through the centre of the base, and do not stop until the other end is reached. See that there is no sudden change of direction in the curve.

The second and third curves shown at C are modifications of the ogee. In the second, the upper end of the upper segment, and the lower end of the lower segment, are drawn fuller than the other parts; while just the reverse is true in the third. But observe, that as in the ogee, so in these curves, there is no sudden change of direction. In drawing these curves, proceed as with the ogee.

To draw the fourth form in C, which is a union of straight and curved lines, first, draw a vertical line. Divide this line into four equal parts, and, through the central point of division, draw a horizontal line equal to one-fourth of the vertical line. Draw the curves and straight lines as shown. Observe that the straight lines so run into the curves, that it is impossible to tell where the former terminate, and the latter begin. This is as it should be, if we would have our lines beautiful. Were the points of junction easily distinguishable, the effect would be much less pleasing. Let this principle of beauty be remembered.

In D we have other varieties of the reversed hand-curve. The base is divided into thirds instead of halves. To draw these curves, proceed as with the ogee. Draw the whole line without stopping, and make no abrupt change in its direction.

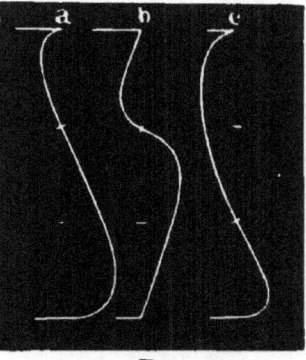

It will thus be seen how easy it is to produce a great variety of reversed curves by changing the divisions of the base line, or by modifying the curvature of the separate segments. Pupils should frequently draw these curves and modifications of them. The contour of articles of glassware, crockery, &c., is usually composed of curves like these.

Reversed Curves applied to Vases, &c.

A VASE. — Draw a vertical line of any given length, and divide it into three equal parts. Through the extremities of this line, draw two horizontal lines equal to one-third of the vertical line, and extending equally on either side of the vertical line. These horizontal lines will give the width of the top and bottom of the vase.

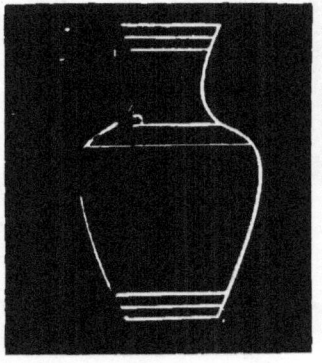

Join the ends of the horizontal lines, giving an oblong as the geometrical base for the design. Divide the last vertical lines into thirds, and, through the upper divisions, draw the curve *b* in D. The breadth, of course, might have been taken less or greater.

A Vase.

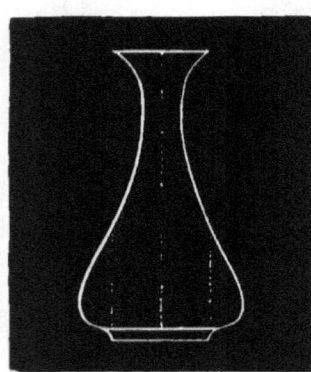

Draw the vertical lines, and divide them as in the last exercise. Use curve *c* in D for the contour of the vase. Add the horizontal line for the bottom.

Require the pupils, using the same curve, to draw other vases of different widths.

A Pitcher.

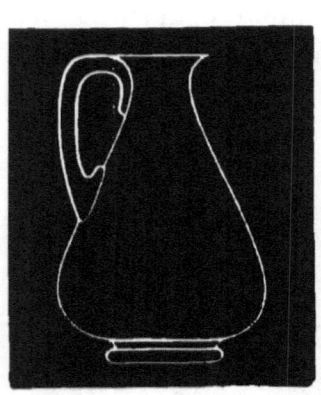

Draw the vertical lines, and divide them as in the last two exercises. For the contour of the pitcher, use curve *a* in D. Add the handle and the bottom. Any other proportions than thirds may be taken, provided the resulting design is graceful, which will depend largely on the drawing of the curves.

It will thus be seen how simple is the mode of designing the contours of vases, articles of table-ware,

&c. Encourage the pupils to exercise their invention in devising new forms.

There is yet room enough for the improvement of common crockery. This improvement must begin with the elevation of the public taste, making ugly things unsalable. With the demand for better work will come the better work. Usually it costs no more to make an object beautiful than to make it homely, provided always that the workman has taste, which his education should give him. If the workman is not able to distinguish between what is beautiful and what is not, it is impossible for him, except by mere accident, to make any thing beautiful. It would be a blind man seeing. Hence the necessity of educating the people, as a whole, to a just appreciation of pure form, which lies at the foundation of all beauty. When the form itself is ungraceful, fine finish and enrichment must always fail to make an object beautiful.

Abstract Curves.—Junction of Curvéd and Straight Lines.

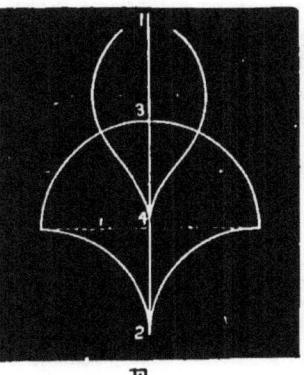

E.

When curves represent nothing, as in this exercise, they are called abstract. The drawing of them is excellent practice. This lesson (E) illustrates another important principle in the treatment of lines, if we would have them beautiful. We have here the side junction of a curved line with a straight line. It will be seen that each curved line, as it approaches the vertical line, changes its direction less and less; and, when it unites with the vertical line, it takes the same direction, and so would not cut across it, however far continued. This is called the tangential union of a curved line with a straight line. The effect is much more pleasing than a secant union, as at F, where the curved line, if continued, would cut directly across the straight line. Study the difference in effect between these two forms of side-junction of straight and curved lines.

In drawing E. draw the vertical line 1 2 of any given length, and divide it into three equal parts. Make the horizontal line equal to two parts. Draw the quadrants springing from 2, and joining the ends of the horizontal line. Draw the compound curves from 4, and then the semi-circle from the upper ends of the quadrants.

Details of Ornament.

Draw a vertical line, and divide it into two equal parts. Make the upper horizontal line, which should be drawn next, equal to the vertical line. All of the curves are compound, except the two shortest. Modify the figure by reversing a part or all of the curves.

Compound Curves.

Card-Exercise XXVII.

Simple Abstract Curves Balanced.

Draw 1 2, and divide it into thirds. Through the lower point of division, draw 3 4, equal to two-thirds of 1 2. Draw the dotted boundary-line, and then the other horizontal lines. Draw the curves which unite with the vertical line at its sides; see that they balance each other. Erase the dotted lines.

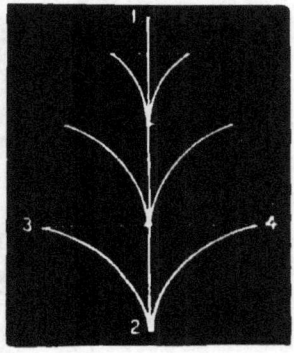

The lines in this exercise might be used as the construction-lines for drawing a leaf; the stem being at 2, the tip at 1, while the veins would follow the balanced curves. (See card-exercise 10, second series.)

Balanced Abstract Curves.

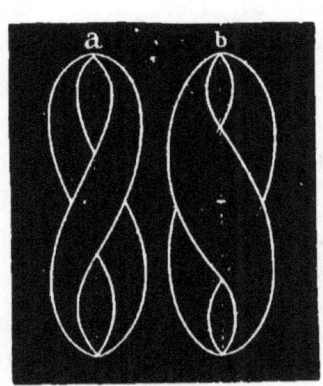

For *a*, divide the base-line into thirds; for *b*, into fourths. Beginning at the top, draw through the upper divisions first, and draw from one end of the base-line to the other without stopping. See that the curves on one side of the base-line are like those on the opposite side. Erase the dotted lines.

Card-Exercise XXVIII.

Abstract Curves. — Junction of Curved and Straight Lines.

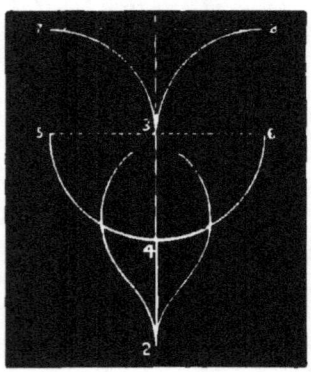

First draw the central line 1 2, and divide it into thirds. Then draw 5 6 and 7 8 through the upper division and the top, making them equal to two-thirds of 1 2. The curves from 7 and 8 are quadrants, that is, quarters of a circle. The curve 5 4 6 is a semi-circle: those springing from 2 are compound curves.

Details of Ornament.

Form a. — Draw a vertical line. Through the first division, from the top, draw a horizontal line twice the length of the first line. Divide this into four equal parts. The width of the flower at the top is two-thirds of the vertical line.

Form b. — Draw a vertical line, and divide it into two equal parts. Through the centre of the vertical line, draw a horizontal line twice as long. Join the extremities of these lines, thus forming a lozenge-shape, indicated by the dotted lines. The curves on the horizontal and vertical lines are compound; the others simple.

Card-Exercise XXIX.

Compound Abstract Curves Balanced.

Draw 1 2, and divide it into thirds. Make 5 6 and 7 8 each equal to two-thirds of 1 2. Draw the lines from 5 and 6 first, and then those ending near 1; next, those from 7 and 8; lastly, those ending near 7 and 8.

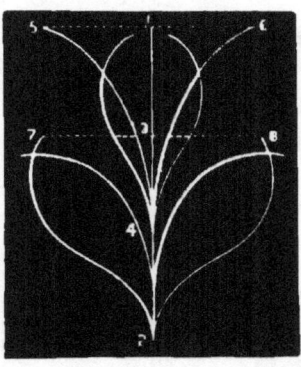

As a rule, draw the long lines first, — those on the left before those on the right; those at the top before those at the bottom.

Card-Exercise XXX.

A Vase. — Reversed Curves Applied.

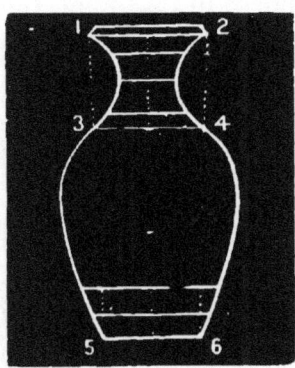

Draw a vertical line, and divide it into thirds. Make 1 2 one-third of the vertical line, 5 6 a little less. Draw 1 5, 2 6. Through the upper division of the central line, draw 3 4. Through points 3 and 4, draw the curves forming the outline of the vase. The line forming the lip of the vase is to be added above 1 2. Draw the parallel lines in the positions shown, the upper one at the bottom through the centre of the lower third of the vase, and the shortest one on the neck through the centre of the upper third. Horizontal lines may be placed on different parts of a vase, but not at its centre. Variety may be obtained by placing them near together in one part, and the reverse in another.

It will be seem that the contour of this vase is a slight modification of the curves at D, in the early part of this chapter.

Card-Exercise XXXI.

Flower-Form. — Flower-Vase. — Ogee Curve Applied.

Form *a*. — For this flower-form, begin by drawing a vertical line of any given length. Divide it into four equal parts. Through the upper end of the vertical line, draw a horizontal line of the same length, one-half to the right, one-half to the left. Draw straight lines, uniting the ends of the horizontal lines to the lower end of the vertical line. Halve these last lines, and, through each point of division, draw a compound curve, with the ogee character, the upper segment tending inwards, the lower segment outwards. Draw the inner compound curves from the ends of the horizontal lines to the lower division in the vertical line. Add the short upper curves.

Form *b*. — Draw a vertical line, of any given length, and divide it into four equal parts. Make the top of the glass one part wide; the greatest width of the bowl

COMPOUND CURVES.

one part and a half, and one part and a half from the top. Make the width of the bottom same as greatest width of the bowl.

FORM *c*.—Draw a vertical line, of any given length, and divide it into thirds. Through the lower end of the vertical line, draw a horizontal line, making it equal to two-thirds of the vertical line. From the upper division of the vertical line, draw oblique straight lines to the ends of the horizontal line. On these oblique lines, draw the compound curves forming the bell. Add the handle on the upper third of the vertical line, and the tongue below.

CARD-EXERCISE XXXII.

Four-pointed Star and Squares.

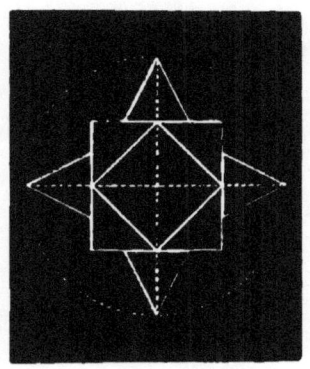

Draw two straight lines of any given equal length, one horizontal, one vertical, and bisecting each other. Divide each half of each line into two equal parts. Through the points of division, draw lines forming a square, having its sides parallel with the first lines drawn. Connect the same points of division by oblique lines, forming a second square within the first. Divide the sides of the first square into three equal parts. On the central part, draw isosceles triangles, with their apexes at the ends of the first two lines drawn. Add the circle, which will test the accuracy of the drawing.

Card-Exercise XXXIII.

Vase.—The Ogee Curve Applied.

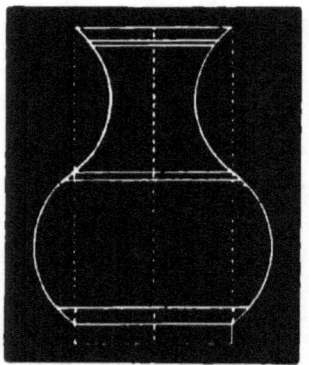

Draw a vertical line. Quarter it. Through each end, draw a horizontal line one-half as long as the vertical line, and extending to the same distance on either side. Draw two other vertical lines, uniting the ends of the horizontal lines. Divide the last-drawn vertical lines into two equal parts; and, through each point of division, draw a compound curve, having the ogee character, to connect the end of the horizontal line above with the end of the horizontal line below. Make the altitude of the curve at the centre of each segment equal to one-half of one-fourth of the vase's height. Add the bottom, making the width of that and the width of the bands equal to one-fourth of one-fourth of the vase's height.

Card-Exercise XXXIV.

A Rosette composed of a Circle and Segments of Circles.

Draw a square, its diameters and diagonals. Divide the diameters into six equal parts, and, through the inner points of division, draw a circle. Draw oblique lines to form the other diagonals of the four small squares. Divide these last lines into three equal parts. Using the last lines as bases, draw on their inner sides simple curves to connect the ends of the diameters of the large square, and with their greatest altitude at the centre, and touching the circumference of the circle already drawn. From the angles of the large square, and from the ends of its diameters, draw simple curves to the points of division on the diagonals of the small squares.

Card-Exercise XXXV.

The American Flag.

Draw the staff and outline of the flag, making the latter slightly oblong. Divide the height of the flag into thirteen equal parts; from the points of division, draw horizontal lines, thus forming the bars or stripes. Let the field for the stars take up seven of the stripes in depth. On this field draw thirteen stars.

The number of original States was thirteen. There are always thirteen stripes on our national flag, representing the thirteen original States; but a new star is always added for every new State admitted into the Union. The copy has thirteen stars, for the thirteen original States. Have the pupil draw the flag with as many stars as there are States at this time.

Card-Exercise XXXVI.
Eight-pointed Star.

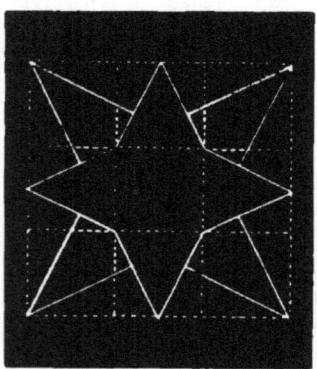

Draw a square, and subdivide it into nine equal squares. Divide each side of the original square into two equal parts. Draw an oblique line from the central point of division on each side to each angle on the opposite side of the square. It will be seen that the smaller squares serve to test the accuracy of the work.

Card-Exercise XXXVII.
A House.

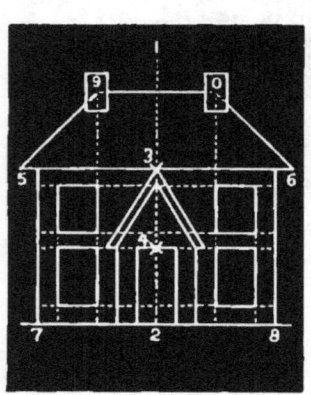

Draw a vertical line, 1 2, of any given length, and divide it into thirds. Through the upper point of division, draw a horizontal line, 5 6, of the same length as the vertical line. Draw a similar horizontal line through the lower end of the vertical line. Draw vertical lines connecting the ends of the horizontal lines. Divide line 7 8 into four equal parts. From the right and left points

COMPOUND CURVES.

of division, draw upwards two vertical lines of the same length as 1 2, the first vertical line drawn. Draw a horizontal line connecting the upper ends of these two vertical lines, and forming the ridge of the house. From the ends of the horizontal line, draw oblique lines for the roof. See that the roof projects somewhat. Add the chimneys, 9, 0. Make the windows and door two-thirds as wide as one-fourth of the width of the house. Make the height of the upper windows a little less than the height of the lower windows. Finish.

Card-Exercise XXXVIII.

A Jug. — The Ellipse Applied.

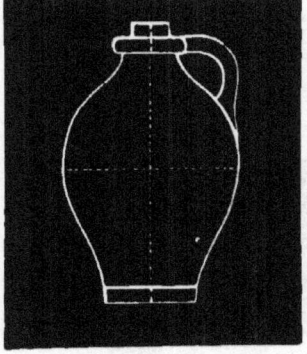

Draw a vertical line of any given length. Through its centre, draw a horizontal line two-thirds as long, and projecting to the same distance on each side of the vertical line. Using these two lines as the diameters of an ellipse, draw the ellipse. Modify the ellipse to form the body of the jug; add the handle and cork.

Card-Exercise XXXIX.

A Vase. — Variations of the Ogee Curve.

Draw 1 2, and divide it into thirds. Draw 3 4, equal to two-thirds of the vertical line. Draw oblique lines, joining the ends of the horizontal line with the

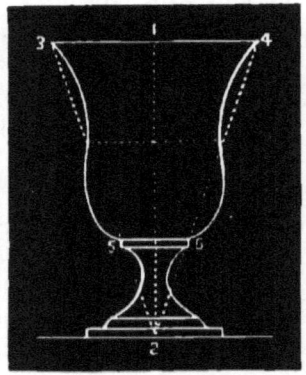

bottom of the vertical line. Through the points of division on the vertical line, draw horizontal lines, to touch the oblique lines. Draw the remaining horizontal lines: make the base nearly equal to one-half of the height of the vase. Finish by carefully drawing the compound curves which form the contour of the vase. See that the curves make a beautiful junction with the straight lines at 5 6. The upper curves are a slight variation of the ogee, the lower segments being a trifle fuller than the upper ones.

Card-Exercise XL.

The British Flag.

Draw the staff of the flag first. The upper and lower lines of the flag are composed of compound curves, and the edge farthest from the staff is a vertical line. The union-jack occupies about one-fourth of the whole flag (seen in this view). Draw the form at the upper right-hand corner in which the bars forming the jack are placed. Its diagonals and diameters, with two parallel lines on each side, will give the outline of the bars. The diameters and diagonals to be then erased.

Compound Curves.

Card-Exercise XLI.

Overlying Forms.

Draw a vertical line, of any given length. Divide it into six equal parts. Through the central point of division, draw a horizontal line two-thirds as long as the vertical line, and divide it into six equal parts. On these two lines as diameters, construct an oblong, and draw its diagonals. With the ends and  sides of the oblongs as bases, draw simple curves through the outer points of division on the diameters. Draw similar curves, connecting the adjacent ends of the diameters and tending inwards. Draw other similar curves, tending inwards, and connecting the adjacent points of intersection of the last curves with the diagonals. Erase the dotted lines, and those parts of the first series of curves which pass under the second series. The width of the border is one-half of one-sixth of the vertical diameter.

Card-Exercise XLII.

Dog-tooth Moulding. — Interlacing Forms.

Draw two squares, lying side by side horizontally, but not quite touching each other. Draw the diameters and diagonals; add the slight indentations at the ends of each diameter. Erase the diameters, and

draw a parallel line above and below, as in other mouldings.

Draw a square and its diagonals. Cut off a trifle from the ends of each diagonal. On each side of what remains of each diagonal, as a base, draw a simple curve, having an altitude equal to one-third of the semi-diagonal. Inside these curves, draw similar parallel curves. Draw each band full, and then erase the portions which do not show because of the interlacing.

Draw a second square and its diagonals. On each side of the square, draw a semi-circle, joining the corners of the square, and passing through its centre. Then draw an inscribed circle. Inside these curves, draw similar parallel curves, except that the curves parallel with the semi-circles must be drawn on a part of each semi-diagonal as a base. Draw the bands full, and then erase the parts which do not show because of the interlacing.

QUESTIONS. — What is a reversed curve? Its segments? What may be the character of each segment? What is meant by a band-curve? Describe the ogee. How should the reversed curve change its direction to be beautiful? How should it be drawn? How should a curve run into a straight line for the junction to be beautiful? How can the shape of the reversed curve be readily modified? For what is this curve much used? What are abstract curves? What is the tangential union of straight and curved lines? What of it? What lines should be drawn first, — the long, or short? the left, or right? How are interlacing forms best drawn?

CHAPTER VI.

PRACTICAL DESIGN.

GEOMETRICAL PATTERNS, ILLUSTRATING SYMMETRY, REPETITION, BALANCE, REPOSE, BREADTH.

Some of the previous exercises illustrate fundamental principles of practical design. But as there were, at the beginning, so many other things to teach, it was thought best to defer all systematic explanation of design until the pupils came to the second series of cards. It will not be difficult, even for quite young children, to understand the principles which will now be discussed and illustrated. A knowledge of these principles will give an additional charm to drawing: it will enable the pupils to begin, at once and intelligently, to make original designs. Though these principles of design will be found so easy of comprehension, like the fundamental principles of arithmetic, yet their applications can be made sufficiently difficult to test the ability of the hand, eye, and taste, which have been thoroughly trained.

Perhaps it may be thought, at the first glance, that some of the drawings on the cards in this series are too elaborate for young children; but, if they have been fairly instructed in the preceding exercises, and are now made acquainted with the principles of design which these drawings illustrate, their execution will be found none too difficult, if we would secure good results. What may at first strike the pupils as intricate will, even to them, appear simple enough, the moment the principles controlling the design are understood.

SYMMETRICAL ARRANGEMENT ABOUT A CENTRE.

First of all, teach the pupils what is meant by symmetrical arrangement about a centre. This requires, that whatever is done in one part of a design shall be done in all corresponding parts. When a line, for example, is added in one part, or removed, lines in all other parts having the same relation to the centre of the design must be added or removed. The moment a line is added without its corresponding lines, the design is spoiled, because there is a violation of the principle of symmetry. There must be no superfluous lines. On the other hand, every time a new line, with all its corresponding lines, is added to a figure, a new design is produced. The new design may be more beautiful, or it may be less beautiful, than the old. Thus it frequently happens, that an elaborate design contains in itself, marking the different stages of its growth, quite a number of less elaborate designs, which are readily discovered by analysis. The arrangement of the parts, many or few, of every design, is always determined by fixed law: nothing is left to mere chance. Illustration will make this clear. There is no one so young or so stupid as not to take greater delight in his drawing when he has been once made acquainted with the law controlling the structure of the design.

A Cross, illustrating Symmetrical Arrangement about the Centre of a Square.

Draw the figure at A on the blackboard, requiring the pupils to draw it with you, on their slates. The exercise may be conducted somewhat after the following manner: —

Teacher. — Draw a square on its diagonals; add its diameters, all in light line.
Pupils. — We have.

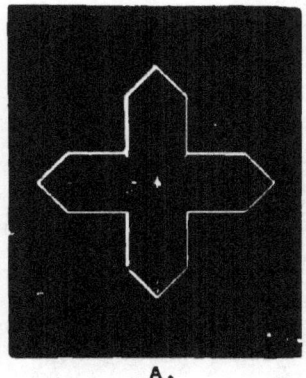

A.

The pupils will indicate that they have completed their work by raising their hands, or in any other way the teacher may direct.

Teacher. — Mark the centre of the upper half of the upper left-hand side, as you see that I have.
Pupils. — We have.

This will give what is seen at *a* in B. Having explained what is meant by symmetrical arrangement about a centre, proceed: —

Teacher. — How many other marks can be made, having just the same position with reference to the centre of the square?
Pupils. — Seven.

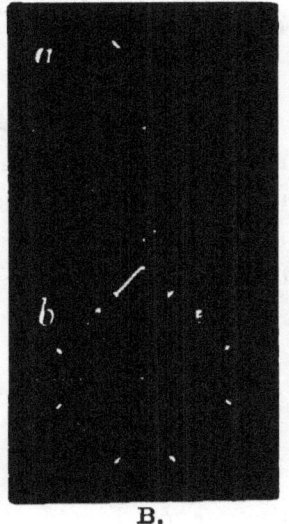

B.

Though this may be their first lesson in symmetry, some of the pupils, nevertheless, will give the correct answer.

Teacher. — Make the seven additional marks.
Pupils. — We have.

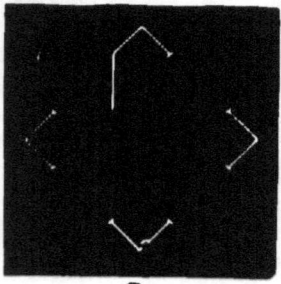
C.

This will give what is seen at *b* in B, less the heavy oblique line.

Teacher. — Now draw the upper quarter of the upper left-hand side of the square, in heavy line, as you see I have drawn it.
Pupils. — We have.

This will give the result which is seen at *b* in B.

Teacher. — How many other lines, having the same relation to the centre of the square, can be drawn?
Pupils. — Seven.

That each pupil may have a fair opportunity to consider the question, insist, if not always, yet certainly at times, that no oral response be given by any one until a signal from yourself. The great art of putting questions properly, is one of the first things a teacher should learn. But to proceed: —

Teacher. — Draw the seven additional heavy lines.
Pupils. — We have.

This will give the result which is seen at C, less the vertical line.

Teacher. — From the lower end of the upper left-

PRACTICAL DESIGN. 109

hand heavy line, draw a vertical line parallel with the diagonal of the square and to the centre of the nearest semi-diameter. Observe me.

Pupils. — We have drawn it.

This will give the result which is seen at C.

Teacher. — How many corresponding lines does symmetry now require to be drawn?
Pupils. — Seven.
Teacher. — Draw them, and erase the construction lines.
Pupils. — We have, and it gives us a cross.

The result may be seen at A, less the construction lines. It is a simple combination of straight lines; but, because the lines are symmetrically arranged about a centre, the effect is pleasing.

A Rosette, illustrating Symmetrical Arrangement about the Centre of a Square.

Here, at D, we have another simple illustration of the principle of symmetrical arrangement about the centre of a square. The form may be used for a blackboard lesson in this wise: —

D.

Teacher. — Draw a square on its diameters; add the diagonals. Thus. From the upper end of the upper left-hand semi-diagonal, mark off one-eighth of its length. Thus. What does symmetry now require?

Pupils. — That the same length be marked off

from each end of the diagonals. We have marked it off.

Teacher. — From the upper end of the vertical diameter, mark off one-eighth of the length of its half. Thus. What does symmetry now require?

Pupils. — That the same length be marked off from each end of the diameters.

Teacher. — On the upper side of the upper left-hand semi-diagonal, as a base, beginning at the point of division, and terminating at the centre of the square, draw a simple curve. Thus. (Seen at *a* in E.)

Pupils. — We have drawn it.

Teacher. — What does symmetry now require?

Pupils. — That seven other similar lines be drawn, having the same relation to the centre of the square.

Teacher. — Draw them accordingly.

E.

Pupils. — We have. (Seen at *b* in E, less the short curve.)

Teacher. — From the upper point of division on the vertical diameter, draw a curve tending from the diameter, to touch the first drawn curve a little below its centre. Thus. (Seen at *b* in E.) What does symmetry now require?

Pupils. — That seven other similar curves be drawn in the seven corresponding parts of the square.

Teacher. — Draw them accordingly; then draw the sides of the square in heavy line, and erase the diameters and diagonals.

The final result is shown at D, less the dotted lines.

At times require the pupils, when working in this way, to make their answers complete statements, as explained in chapter two.

From the illustrations which have now been given, it will be seen that this principle of design — symmetrical arrangement about a centre — is so easily comprehended and so easily observed in practice, that it can be taught to young children almost at the outset of their drawing career. You will often find, in giving lessons from the blackboard, as just illustrated, that attention paid to this principle will greatly abridge your labor, while the pupils will manifest increased delight in their work. You will often find, also, that attention paid to this principle will be of service in giving dictation lessons. Again, pupils, when drawing from copies on the cards, will be greatly helped by a knowledge of this principle of symmetrical arrangement. A design, which would otherwise appear to them intricate and difficult of execution, is discovered to be quite the reverse when viewed in the light of this knowledge. It is seen, that if two or three different lines can be drawn, in themselves simple, then the whole design, which is but a combination of these two or three lines, can be quite as readily drawn. Symmetry is, indeed, the key which unlocks many of the secrets of design and drawing.

See that your pupils, when they examine the drawings on their slates, do not look at them obliquely; for, if they look at them thus, the drawings are distorted, and they fail to see them as they are, and hence cannot tell whether they are correct or not. The eye must be directly opposite the centre of the drawing.

Symmetrical Arrangement about the Centre of an Equilateral Triangle.

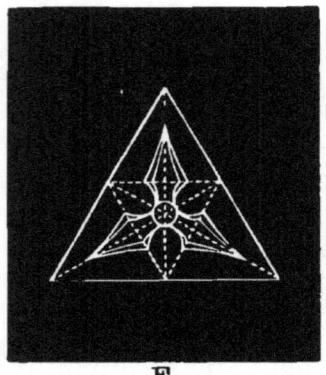

F.

Instead of a square for the geometrical basis, take, this time, an equilateral triangle. Having drawn the triangle, bisect each side; from each point of bisection, draw a straight line to the opposite angle. The lines will cross each other at the same point, — the centre of the triangle. This will give six right triangles symmetrically arranged around the centre of the large triangle. Connect the centres of the sides of the original triangle by straight lines. Add the curved lines as at F.

G.

You might stop here, for the design is symmetrically perfect. Take another step, however, and add the curved lines as at G; also the outer straight lines parallel with the inner triangle. Erase the dotted lines. Thus you obtain a new design. You might, in a similar manner, take another step, and obtain a third design; and so on.

This illustrates what was meant when it was said that frequently an elaborate design contains in itself, marking the different stages of its growth, quite a number of less elaborate designs, which are readily discovered by analysis. Thus, the design at G contains the one at F, while further analysis would disclose others contained in F.

Symmetrical Arrangement about the Centre of a Circle.

H.

Draw a circle on its horizontal and vertical diameters; add two other diameters, dividing the circle into eight equal parts. From each end of the horizontal and vertical diameters, mark off one-sixth part of a semi-diameter. Starting from the left-hand point of division on the horizontal diameter, draw an ogee curve, terminating in the right-hand point of division. Repeat this curve as many times as required by symmetry. Add the fan-like forms on the other diameters, as seen at H.

If you give the form at H or G as a blackboard lesson, draw only one curve at a time, and let the pupils add the others according to the law of symmetry. If you give either as a dictation exercise, in which case you can easily fix upon the measurements, proceed in a similar manner, dictating one line, and requiring the pupils to add all the corresponding lines.

For other examples of symmetrical arrangement about the centre of a square, circle, hexagon, pentagon, and triangle, see the designs A, B, C, D, E, in the next chapter, where a conventional leaf is the principal element.

From time to time review the pupils in the principles explained in the first series of cards.

SYMMETRICAL ARRANGEMENT ON AN AXIS. — BALANCE OF PARTS.

The next thing it will be well for you to teach your pupils, is the meaning of symmetrical arrangement on an axis, that is, on a straight line. This requires balance of parts, — that the right half of a design should be like the left. When you have symmetrical arrangement about a centre, any straight line drawn through the centre will divide the design into two equal parts, which will always be just alike. But when you have symmetrical arrangement on an axis, the design must be divided on the axis, otherwise the parts will be unequal and unlike. In the first case, if the symmetrical arrangement is about the centre of a triangle, then a line once drawn must be thrice or six times drawn; if about the centre of a square, it must be four or eight times drawn; if about the centre of a pentagon, it must be five or ten times drawn; if about the centre of a circle, it must be drawn as many times as the circle can be divided into parts, like the one in which the first line of the given kind is drawn. But in the second case, when a line has been once drawn, it can only be drawn a second time, the second line balancing the first.

The larger part of the exercises in the second series of cards illustrate symmetrical arrangement on an axis. Each of the buds, for example, at *b*, in Exercise I., second series, is symmetrically arranged on the vertical diameter of the square which encloses it. Observe that this line alone will divide the bud into two equal parts; not so with the rosettes at *a*, in the same exercise. In Exercise II., second series, we have a good illustration of symmetrical arrangement on an axis, combined with symmetrical arrangement about a centre. Each separate bud is symmetrically arranged on an axis; while the design, considered as a whole, is symmetrically arranged about a centre.

REPETITION, HORIZONTAL AND VERTICAL.

When you have made a design after the manner of those we have been considering, you can, for many practical purposes, take the design as a unit, and repeat it an indefinite number of times, either in a horizontal or in a vertical direction. This is done in architectural decoration, in the designs for carpets, and all kinds of ornamented textile fabrics, and in covering surfaces generally. Often a few lines are added for the purpose of uniting the repetitions of the original unit. Sometimes the unit itself may have little or no beauty; yet, if repeated vertically or horizontally, the result becomes very pleasing. Exercise I., second series, illustrates horizontal repetition; while Exercise V., same series, illustrates vertical repetition.

REPOSE AND BREADTH.

When the eye falls upon a design, and is content to remain where it falls, that quality in the design which produces this quieting, agreeable effect, is denominated repose. If there is a lack of repose, then the eye feels impelled to wander over the design, or off it, finding no rest. The sensation is decidedly uncomfortable. Thus there is no rest for the eye when looking upon a carpet which has only straight, parallel lines; but it feels impelled to follow the lines hither and thither across the room. Introduce other straight lines crossing the first, and the wandering tendency of the eye is diminished. If, instead of parallel straight lines, the design consists only of concentric circles, then the tendency of the eye is to move round and round, following the circles. This tendency is diminished, or ceases altogether, when the circles are crossed by a few straight lines.

In order, then, that a design may have repose, there must be no long, uninterrupted lines (comparatively long for the size of the design), either straight or curved; but there must be a due combination of lines, straight and curved, vertical, horizontal, and oblique. While attempting, however, to secure repose, avoid an undue multiplication of lines; for too many lines confuse and destroy the effect. See, too, that there are some lines much more prominent than others, — more prominent by reason of their length, and not of their size, at this stage of drawing; for, if there are not such lines, the design will lack breadth, — it will not please at the first glance, and so will not invite further inspection, however much its details might please when carefully studied.

PRINCIPLES OF DESIGN DERIVED FROM NATURE.

The principles of practical design are derived from a study of nature, especially from a study of the vegetable world, — of flowers, fruits, leaves, vines. When we examine flowers, we find their petals symmetrically arranged about centres; while each petal, considered by itself, is symmetrically arranged on an axis. When we examine leaves, we find that each is symmetrically arranged on an axis, which is the midrib in leaves having a single lobe. This, of course, is only true of the general form of the leaf, not of its details. Then, too, an occasional leaf violates, in a more decided manner, the law of symmetrical development, as does the elm-leaf, with its oblique base, while some flowers are so irregular they cannot be classified, and are, therefore, called anomalous. Indeed, nature uses all forms; but, while she does this, she shows a decided preference for the symmetrical.

The law of symmetrical development on an axis holds true, in the main, of animal forms. Thus the human figure consists of two equal, similar parts, the one balancing the other.

What is true of those products of nature that can be studied by the unaided eye, is found to be equally true of the microscopic world. Examine snow-flakes under a microscope: the crystals exhibit a great variety of charming forms; yet each is symmetrically arranged, almost with geometrical precision, about a centre.

In distributing her lines of growth, nature is also observant of breadth and repose. Examine the structure of a leaf. There are the ribs, or only a midrib, as the case may be, strongly defined; then come the veins, less strongly defined, and taking a different direction; and then a further subdivision of the surface by veinlets, and even finer tracings. What is true of the leaf, is true of the tree as a whole. Standing beneath the tree, and looking up, we behold the large branches into which the trunk divides; then follow the minor subdivisions, down to twigs and leaves: this is a combination which gives at once breadth as the result of the general features, and repose as the result of the added minor features. All is harmonious and agreeable.

Require your pupils to bring into school leaves, flowers, shells, insects, that their forms may be studied in connection with the lessons in drawing. Require them to bring the greatest variety possible, not only to be compared with one another, but to be drawn rapidly. The study of art and the study of nature will thus go hand in hand, and each will help the other. When there is an intimate connection between two things, they can both be learned together, as easily as either can be learned alone, and oftentimes with less effort.

ORIGINAL DESIGNS BY THE PUPILS.

Guided by the principles which have now been explained, the pupils will be able to construct original geometrical designs, with both ease and understanding. They should be permitted, therefore (for hardly more than permission will be necessary), to exercise their inventive powers in this manner. They can use the construction lines employed in the copies, and draw upon these, as a basis, lines which will give new combinations; or they can modify the construction lines themselves, by dividing them differently, or by adding new ones. In devising their new geometrical designs, they will be limited by the alphabet of lines, straight and curved: they will have nothing to do with natural forms, except to follow the general principles deduced from a study of these forms. Farther along, the direct use to be made of these forms in practical design will be explained and illustrated.

Inspect the designs originated by the pupils; criticise kindly; commend whenever you can. Do not judge their productions by a high standard. Too many lines will be, perhaps, their greatest defect at first, destroying the breadth of the design.

With the hints here given, teachers can make the succeeding exercises very agreeable work for their pupils. When drawing is taught, not simply as an exercise of copying forms, but as suggested in this chapter, it becomes an exercise which cultivates, in a wonderful degree, the observing and inventive powers on the one hand, while it imparts skill in the visible expression of knowledge on the other.

Card-Exercise I.

Horizontal Repetition. — Rosettes and Buds.

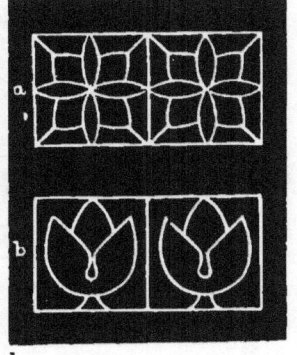

Form *a.* — Draw two squares touching each other horizontally. Draw the diameters and diagonals; divide the diagonals into six equal parts. On the diameters, draw compound curves passing through the centres of the squares; add the leaflets which terminate at the outer points of division on the diagonals.

Form *b.* — Draw two squares as in the last. Draw their diagonals and vertical diameters; divide the upper halves of the former, and the lower halves of the latter, into thirds. From the upper points of division on the diagonals, draw curves through the lower points of division on the diameters. Finish as indicated.

VARIATIONS. — Make a horizontal moulding composed of alternate squares of *a* and *b*, — a rosette, then a bud, — and not less than three of each. Again, make a vertical moulding by repeating a single square and bud of *b*.

Card-Exercise II.

Buds repeated around a Centre.

Draw two straight lines, one vertical, the other horizontal, bisecting each other. Divide each into six equal parts. Through the inner points of division, draw two horizontal and two vertical lines of the same length as the first. Unite the ends of these lines by two horizontal and two vertical lines, thus forming five equal squares. Draw the diagonals of the four outer squares, indicating the position of the leaves of the buds to be drawn. On each diameter of the inner square, draw two compound curves, and, in each of the other squares, draw a bud similar to *b* in the last exercise.

Card-Exercise III.

Coffee-Pot.

Draw the central vertical line, and divide it into fourths; add the other vertical lines, making the width of the oblong equal to one-half of its height. Through the centre of the upper fourth of the vertical line and extending the same distance on each side, draw a horizontal line, for the top of the coffee-pot, equal to one-half of the width of the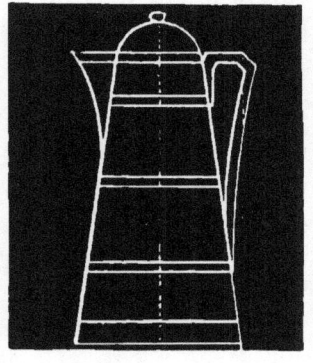
oblong. Draw the outline of the coffee-pot and its lid. Add the spout and handle. The lower band covers one-fourth of the lower fourth of the central line. The other bands are one-half as wide. Finish.

Card-Exercise IV.

Cup and Saucer, Salver, and Muffin-Dish.

Form *a.* — For the cup and saucer, draw a vertical line, and divide it into thirds. Through the lower point of division, draw a horizontal line, and another immediately below, twice as long as the vertical line. For the top of the cup (not including the handle), draw a horizontal line one-third longer than the vertical line. Finish.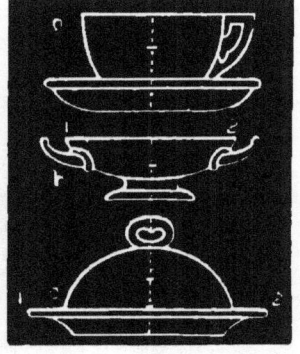

FORM *b*. — For the salver, draw a vertical line and halve it. Make 1 2 thrice as long as the vertical line. Draw the long curve of the bowl through the centre of the lower half of the vertical line; add the handles and the base.

FORM *c*. — For the muffin-dish, draw a vertical line, and divide it into four equal parts. Make 1 2 twice the vertical line. The handle occupies the upper fourth of the vertical line. The bowl of the cover is a semi-ellipse, its width rather more than twice its height. Finish.

VARIATIONS. — Change proportions and curves. Thus, make the outline of the saucer in *a* a compound instead of a simple curve. Modify the salver by making 1 2 five times the vertical line, changing nothing else.

CARD-EXERCISE V.

VERTICAL REPETITION.

Guilloche. Leaves and Berries, Conventionalized.

FORM *a*. — On a vertical central line, draw three squares touching each other. Draw the diagonals of each square to find the centre. Around each centre, draw a circle, with a diameter equal to one-third of the diameter of the square. Around these three circles draw the endless curved band. The dotted lines show the parts which are overlapped. The straight lines visible in the running pattern are drawn on the diagonals of the squares. Erase the dotted lines.

This is a sample of the guilloche, an ornamental form which was used by the very earliest historic nations,—as by the Assyrians, for instance. Its peculiar feature consists in the winding and interlacing of the long curves about the circles. This being an historical ornament, and frequently used now, pupils should practise drawing it until they can readily reproduce it from memory.

FORM *b.* — Draw three squares, as in figure *a*, upon the central vertical line; then the long curves, forming the outline of the repeated leaf. Divide the central line of each square into two equal parts, which will give the points on it from which to draw the smaller curves at the base of the leaves. The base of a leaf is the part next the stem. Draw the circles to fill up the spaces left, and the stem upon which leaves and berries are supposed to be growing.

The pupils should now be able to draw quite rapidly, and at the same time make their lines much better than they did at first. See that their pencils are always long and well pointed, otherwise much time will be wasted in attempting to make proper lines. Indeed, you must economize time in every way. Some teachers consume tenfold more time than do others in getting ready for recitations, simply because they have no system. A class of forty pupils should not be more than one minute getting ready for a drawing-lesson, even if all the slates and pencils have to be distributed from the teacher's desk.

Card-Exercise VI.

Horizontal Repetition.

Guilloche.   Leaves and Berries, Conventionalized.

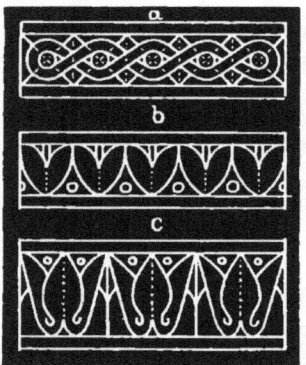

Form *a*. — Draw five squares, touching one another horizontally. Fill them with an endless band, as *a*. in the last card-exercise, was filled. Add the parallel side-lines.

Form *b*. — Draw five squares, touching each other horizontally. Fill them with repeated leaves like those in *b*, last card-exercise. Add the parallel side-lines.

Form *c*. — Draw three squares, touching horizontally. Add their vertical diameters, as construction lines, and then fill the squares with repeated flowers. as shown in the copy.

Card-Exercise VII.

Simple and Compound Abstract Curves, Balanced.

The curves 3 2, and 4 2, those which end nearest 1, and the spirals at the bottom, are all compound. The other two pairs are simple curves. Remember this when drawing the curves.

Draw 1 2, and divide it into four equal parts. Make 3 4, 5 6. equal to three-fourths of 1 2. Draw. first, the longest curves 3 2, 4 2, balancing them equally on each side of the vertical line. Draw the simple curves springing from 5 and 6 to join the curves first drawn. Divide 1 3 into four equal parts, also 1 4. From the central points of division, draw the simple curves to the centre of the vertical line; and, from the points of division nearest 1, draw the tulip-shaped curves terminating at the lowest point of division on the vertical line. Draw the spirals, with their ends joining the ends of a horizontal line drawn two-thirds of one part of the vertical line above 2. See that all the lines run gracefully into one another.

Card-Exercise VIII.

Simple and Compound Abstract Curves, Balanced.

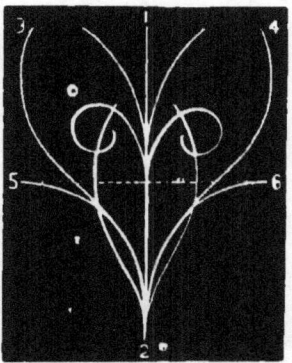

It will be seen that this is a slight variation of the last exercise, the spirals being drawn near the centre of the vertical line, and the long compound curves, terminating near 1, being drawn from the bottom of the vertical line. Draw, first, the lines on the left, and then those on the right to balance.

THE SPIRAL.

There are two varieties of the compound curve called the spiral. In the first variety, the distance between the different spires is the same, as shown at I. This is called the equable spiral, because its breadth increases uniformly with each revolution; it is also called the Archimedean spiral, from a Greek mathematician who lived more than two thousand years ago, and was the first to define its characteristics.

If an upright cylinder, or the round trunk of a tree, be wound with a single layer of cord, and, if a person then take hold of the end of the cord, and unwind it by walking around the cylinder or tree, keeping the cord all the while taut, his path will be that of an equable spiral (I). Each time he makes a revolution he will increase his distance from the cylinder by just the circumference of the cylinder. This may be illustrated, in a small way, by using a round pencil and thread.

To construct the first form of the spiral I, draw two straight lines, o and n a, crossing each other at right-angles. At the point of intersection, draw a small circle to represent the end of a cylinder or pencil. This circle is called the eye of the spiral. Let the spiral begin on the horizontal line at b, cross the vertical line above at 1, the horizontal line (left) at 2, the vertical line below at 3, the horizontal line (right) at 4. The figures 1, 2, 3, 4, indicate that the distance of the spire from the eye of the spiral at those points is, respectively, equal to one-fourth, two-fourths, three-fourths, and the whole, of the circumference of the circle, according to

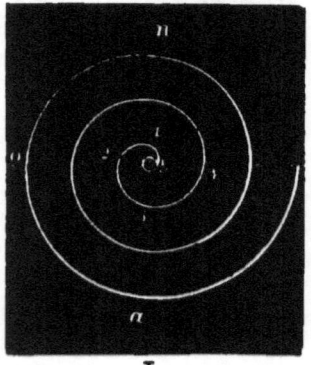

I.

the part of the revolution accomplished. The distance from h to 4 is, therefore, equal to the circumference of the circle, or would be in an actual illustration with a cylinder and cord.

Before beginning to draw the spiral itself, mark on the horizontal and vertical lines the points through which the spiral is to pass. With these points fixed, the curve can be quite readily drawn. When the spiral is finished, the circle may be erased, as that forms no part of it. Draw a spiral without the aid of a circle. After the points have been marked as directed, on the horizontal and vertical lines, other horizontal and vertical lines, similar to those in J, may be drawn through the points to assist in the construction of the spiral, if it is a large one. The spiral should be frequently drawn for practice.

In the second variety, or variable spiral seen at J, the distance between the different spires increases according to some regular proportion, as you leave the centre. This spiral may be illustrated by the unwinding of a cord from a truncated cone, instead of a cylinder, beginning at the smaller end. The person who unwinds the cord will find, that, with each succeeding revolution, he increases more rapidly his distance from the cone, his path being that of the variable spiral (J).

To construct a spiral of this character, draw two straight lines, o and f, crossing each other at right-angles at c. Draw two other straight lines, s and t, crossing each other at right-angles at c also; but make the angle $o\ c\ s$ slightly smaller than the angle $s\ c\ p$. Beginning at any point on the line o to

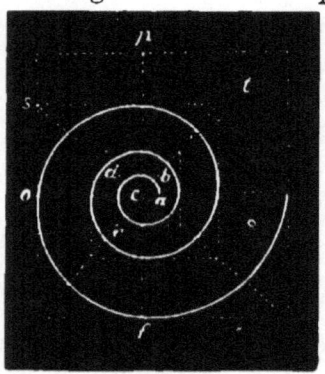

the right of the line $p\ f$, as at a for example, draw the line $a\ b$, parallel with the line $p\ f$. From the point b, where the line $a\ b$ touches the oblique line t, draw $b\ d$ parallel with the line o. From the point d, where the line $b\ d$ touches the oblique line s, draw a vertical line downwards until it touches t at e. Thus proceed with the straight lines. Starting from a, draw the spiral through the points where the vertical and horizontal lines intersect. Erase the dotted construction lines. This spiral might, of course, be drawn in a manner similar to the first.

CARD-EXERCISE IX.

ABSTRACT CURVES, BALANCED SPIRAL FORM.

DRAW the central line 1 2, and divide it into four equal parts. Through the lower point of division, draw the horizontal line 3 4, making it equal to one-half of 1 2. A little below the centre of the upper fourth of the vertical line, draw 5 6, of the same length as 3 4. Divide 3 4 into four equal parts, and 5 6 into six equal parts. Next draw the long curves, starting them from 5 and 6, and finish by adding the small curves.

When drawing the spiral, frequently turn it upside down for examination. Unless guarded against it, the eye easily becomes accustomed to an error in a curve of this character, and at last cannot even detect it. By viewing the curve in all positions, the liability to error is lessened. As a preliminary lesson, it will be well to give the figure with all the minor lines omitted. Afterwards draw it complete.

The spiral is a most difficult curve to draw well; and, as it is of the first consequence in refined drawing, it should therefore be much practised. When one can draw the spiral readily and well, he will have but little trouble with other curves.

VARIATIONS. — Draw the figure in a reversed position. Also draw it twice above a horizontal line passing through 2, and twice below the horizontal line in a reversed position.

Square Rosettes Repeated.

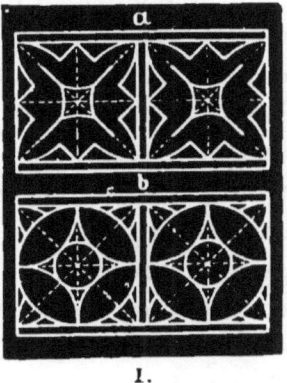

FORM *a*. — Draw a horizontal line; on this construct two equal squares at a slight distance apart. Add their diagonals and diameters. Next draw the short oblique lines, those starting from the ends of the diameters to be drawn parallel with the diagonals. Add the curves, a horizontal line above, and another below.

FORM *b*. — Draw two squares, their diagonals and

diameters, as in form *a*. Inscribe a circle in each square. Make the diameter of the smaller circles a trifle more than one-third of the diameters of the larger circles. Connect the ends of the diameters by simple curves touching the circumference of the inner circles. Add the other curves, and a parallel line above and below. These two forms are for blackboard lessons; or, with a slight change in the descriptions, they can be used for dictation lessons.

GEOMETRICAL DESIGN.

With two or three exceptions, the designs given in this chapter are simply geometrical: they represent neither natural nor artificial objects. But such designs often possess a great degree of beauty. This is especially true when the lines are combined according to principles derived from a study of nature. Geometrical designs have had the indorsement of good taste in all ages. One of the very oldest ornaments, and the one which has been most widely used, is the Greek fret. The Greeks did not originate, they only elaborated, this ornament, which appears to have been used, in some form, by every people, civilized and savage, and in every age, even the remotest, of which we have any record. To-day no ornament is so generally employed: it is employed for decorating architecture, the products of the loom, pottery, indeed, almost every thing.

Yet no ornament could well be more arbitrarily geometrical, as shown by the sample given in Card-Exercise XVII., First Series. It possesses a great degree of repose; it illustrates the continuous growth of the vine; and it can be easily modified for a variety of purposes.

But the countries which have carried geometrical design to the highest pitch of perfection are those where the Mohammedan religion prevails. Forbidden by this religion to make representations of natural objects, the people are thus compelled to rely upon geometrical ornament alone. Moresque design, in particular, shews what exquisite results may be obtained from the use of such limited material, when the lines are combined according to the principles of nature, and yet without any attempt to represent, even remotely, natural forms.

QUESTIONS. — Are the principles of practical design hard to understand? What of their application? What is said of symmetrical arrangement about a centre? Of superfluous lines? Of one design including others? How may the principle of symmetrical arrangement about a centre be used in blackboard lessons? In dictation lessons? How does symmetrical arrangement on an axis differ from symmetrical arrangement about a centre? What is said of repetition? Of its use? What is said of repose? Of breadth? Illustrate each. From what are the principles of practical design derived? What is said of original designs by the pupils? What will be their greatest fault at first? What is said of the guilloche? Why should the spiral be frequently drawn? What is said of geometrical ornament? Of the Mohammedan religion?

CHAPTER VII.

PRACTICAL DESIGN.

LEAVES, BUDS, FLOWERS, CONVENTIONALIZED.

WITH two or three minor exceptions, all of the preceding exercises, which have illustrated principles of design in any way, have consisted of geometrical patterns, or outline representations of artificial objects. We now come to exercises which not only illustrate the principles of design derived from nature, but give conventionalized representations of natural objects. By such representation, it is meant that only the general form is imitated, — that the object is symmetrically drawn; all the minor irregularities being omitted. It is taking the symmetrical before the unsymmetrical, the conventional before the natural; and this accords with sound principles of teaching. Then, too, practical design concerns itself mainly with conventional forms: it seldom attempts (when it does it violates the best taste) an exact representation of nature. Even the highest art, as illustrated in **statuary and painting**, though founded on the imitation of nature, goes farther than mere imitation, and gives a rendering of nature in harmony with the requirements of the most cultivated taste. It is only in the painting of flowers, fruits, and the like, that an exact imitation of nature can be properly attempted.

PRACTICAL DESIGN. 133

The following exercises, A, B, C, D, E, illustrate the power of repetition in numbers on form:

Conventional Leaves symmetrically arranged about the Centre of an Equilateral Triangle. The Power of 3.

Draw an equilateral triangle. Divide each side into halves; and, from each point of division, draw a line to the opposite angle. This will divide the triangle into six equal parts. In each third, draw a conventional leaf, as seen in the copy (A). Add the curves near the centre.

A.

Observe that in making each of these five designs, A, B, C, D, E, we have used the same elements; modifying them just enough to adapt them to the spaces to be filled. The leaf-form which has been used is, of course, a highly conventional one.

Require your pupils to make new designs, by combining, in a similar manner, the leaves and other forms given on the cards. This is one of the first steps in design, putting old forms into new combinations, and will be found exceedingly pleasant and profitable for the pupils.

When a pupil has made a new design on his slate, permit him to reproduce it on the blackboard that the whole school may see it. Then require him to tell how he drew his design, and describe the principles which it illustrates. For the pupils to describe, in a like manner, any of the card-exercises will be very serviceable in giving them a command of language.

134   TEACHERS' MANUAL.

Conventional Leaves symmetrically arranged about the Centre of a Square. The Power of 4.

Draw a square, its diameters and diagonals. Add straight lines connecting the ends of the diameters, and divide them into thirds. In each quarter of the square, draw a conventional leaf, as seen at B. Add the curves near the centre, the outer points of the triangular forms falling on the centres of the semi-diameters. Give this as a blackboard lesson, following the method described in the last chapter.

B.

Conventional Leaves symmetrically arranged about the Centre of a Pentagon. The Power of 5.

C.

Draw a circle on its horizontal and vertical diameters. Erase the horizontal diameter, and divide each half of the circle into five equal parts, making ten equal parts in all. Draw straight lines uniting the alternate points of division; these chords will form the sides of the pentagon. In each fifth of the pentagon, draw a conventional leaf, as seen in the copy (C). Add the curves near the centre.

PRACTICAL DESIGN.

Conventional Leaves symmetrically arranged about the Centre of a Hexagon. The Power of 6.

Draw a circle, and divide it into twelve equal parts,

D.

or sectors, by first dividing it into halves, then into fourths, and then each fourth into thirds. Draw chords connecting every other point of division; these six chords will form the sides of the hexagon. In each sixth of the hexagon, draw a conventional leaf, as seen in the copy (D). Add the curves near the centre of the hexagon. The outer points of the curved triangles terminate at the centres of the semi-diameters of the hexagon.

Conventional Leaves symmetrically arranged about the Centre of a Circle. The Power of 8.

Draw a circle of any given diameter, and divide it into sixteen equal parts, or sectors, first by dividing it into halves, then into fourths, then into eighths, then into sixteenths. Draw a second circle, having the same centre and one-fourth of the diameter of the first. In each eighth of the large circle, draw a conventional leaf, as seen in the

E.

copy (E). Add the curved triangles near the centre. The bases of these triangles are parts of a circle having a diameter equal to one-third of the diameter of the large circle. Analyze this design, and see how simple are its elements.

You will soon find that the pupils who draw best do not always succeed best in original designs. Manual skill and inventive power are not always united in equal proportions; hence it will sometimes happen that a pupil, discouraged by his poor manual execution, will take heart again when he begins original design, because he succeeds so well.

Oblong Rosette of Irregular Conventional Leaves.

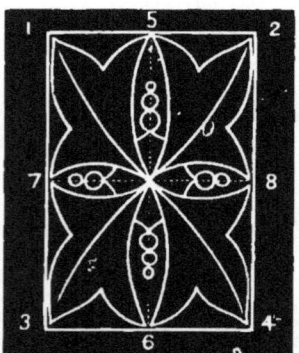

Draw an oblong, making its width two-thirds of its length. This can be done by drawing the vertical line 5 6, dividing it into thirds, and then drawing 1 5 and 5 2, each equal to one of the thirds. In the same manner, draw 3 4. Then add 1 3, 2 4, and draw 7 8 through the centre. On the diameters, draw four ogee curves, drawing each curve from one end of the diameter to the other without stopping. Next draw the slightly curved lines which form the midribs of the leaves, and then the curves forming the lobes. Now draw the two short curves which meet on each semi-diameter at a distance from the centre of the oblong equal to one-half of a short semi-diameter. Where these curves meet, add the berries as seen in the copy, two on each short semi-diameter, and three on each long semi-diameter.

VARIATION. — Draw a square rosette as in the preceding exercise; then add the berries, two on each semi-diameter.

Square Rosette of Irregular Conventional Leaves.

Draw a square and its diameters, and on the latter draw four ogee curves. Next draw the four slightly curved lines which form the midribs of the leaves; then add the curves forming the lobes. Observe that while each leaf is slightly irregular, the four are just alike and symmetrically arranged about the centre of the square.

Lotus-Flower Rosette, Square.

Draw a square, its diagonals and diameters. Parallel with the diagonals, draw the lines which form the outlines of the lily or lotus-flowers. Draw the central leaf of each, and finish according to the copy. This can be easily used for a blackboard lesson.

The lotus-flower is an Egyptian ornament, symbolic of the Nile, source of fruitfulness and plenty.

VARIATION. — Instead of drawing the lotus flowers on the semi-diameters, draw them on the semi-diagonals. Be certain to retain the general characteristic of the flower.

Lotus-Flower Rosette, Oblong.

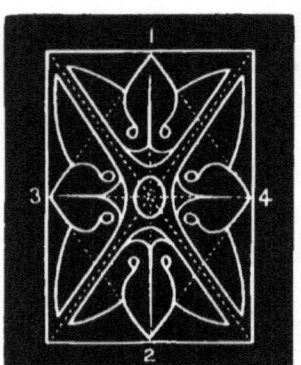

Draw the line 1 2, and divide it into three equal parts. Make the ends of the oblong equal to two-thirds of 1 2. Complete the oblong, and draw its diagonals and short diameter. Connect the ends of the diameters by straight lines, and divide these into fourths. Parallel with the diagonals, draw the lines forming the outlines of the lotus-flowers. Adapt each flower to the space to be filled. This illustrates symmetrical arrangement around the centre of an oblong, and will serve for a blackboard lesson by the teacher.

Blackboard and Charts.

See that you make only a proper use of the blackboard. If you attempt to teach a class wholly from that, making your blackboard copies take the place of all printed copies, then it is absolutely essential that you be able to draw both rapidly and beautifully. The development of taste is one of the chief objects of drawing; that cannot be secured unless the pupils have beautiful copies, models, to look upon and to draw, — the more beautiful the better. Scarcely any one can draw with sufficient rapidity and beauty to provide such copies wholly from the blackboard.

But this is not all. When pupils draw from copies on the blackboard, if the class is large, only a few of the whole number can obtain a good view of the copy, — the few who sit directly in front. For those who sit to the right or to the left, every line, except vertical ones, is foreshortened; that is, the fore or front view does not give the true length of the line, because it is seen obliquely. Thus, a square appears to be an oblong; a circle, an ellipse; all but vertical lines are distorted; forms which may be, in reality, beautiful, lose all their beauty. While, therefore, only a few of those having good vision can see the blackboard copy as it is, others having defective vision cannot see it at all because of the distance. These same grave objections hold against drawing from charts, and so they are forbidden to be used in the best European schools.

Hence it is absolutely essential, if pupils are to make satisfactory progress in drawing, that each one should have beautiful copies placed in his hands; and, in addition to these, there must be beautiful models, and always *clean*, whatever their color, when he comes to draw from the solid. These copies, in the hands of the pupils, greatly diminish the labor of the teacher, and are absolutely essential for developing the taste of the pupils.

As frequently affirmed before this, the blackboard has its important uses. For illustrating principles it should be often employed, since you can then instruct a whole class at once. Rude drawings, rapidly executed, are usually sufficient for this purpose. Then it is essential that you should often draw on the blackboard the figure which the pupils are drawing from their copies; you can thus keep the whole class together, teach all at once how to draw the given figure, spur the indolent to greater activity, and restrain the overhasty. Again, it is sometimes well to give on the blackboard entirely new exercises, thus compelling the pupils to observe carefully every step you take, and to repeat the same after you. Also, for the purpose of verifying dictation exercises, you must use the blackboard: otherwise you must take time to examine the work of each pupil separately, however large the class may be. This limited use of the blackboard, and this only, has the approval of the best experience.

Remember that several things, in the way of discipline, are to be aimed at, and the mode of instruction must be modified accordingly.

Card-Exercise X.

Conventionalized Leaf.

Draw the vertical line 1 2, and divide it into thirds.

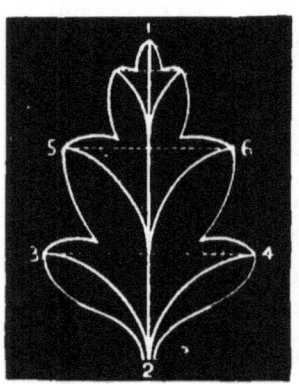

Make 3 4 equal to two-thirds of 1 2. Draw the dotted curves forming the boundary-line, and then the other horizontal lines. Next draw the curves from the ends of the horizontal lines to the vertical line. Finish. This is the conventionalized form of no particular leaf. Observe that it is built on the abstract curves in Card-exercise XXVII., of the First Series. This illustrates symmetrical arrangement on an axis.

Card-Exercise XI.

Crocus Flower, Conventionalized.

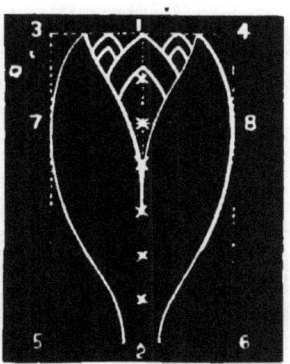

Draw the vertical line 1 2, and divide it into seven equal parts. Draw 3 4 and 5 6 equal to four of these parts. Join the ends of these horizontal lines by vertical lines, and draw 7 8 through the second point of division from the top. Draw the long curves of the flower first, and then the shorter ones. Here we have another illustration of symmetrical arrangement on an axis.

Card-Exercise XII.

Convolvulus Bud, Conventionalized.

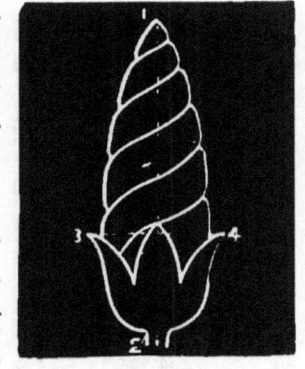

Divide the vertical line 1 2 into thirds. Make 3 4 equal to one-third of 1 2. Sketch lightly an unbroken boundary-line of the whole bud; and then finish as in the copy. Observe that one of the curves passes through the upper point of division on the vertical line. From the necessity of the case, the drawing of the spiral portion of the bud cannot be perfectly symmetrical above the cup.

Card-Exercise XIII.

Sprig of Convolvulus, Conventionalized.

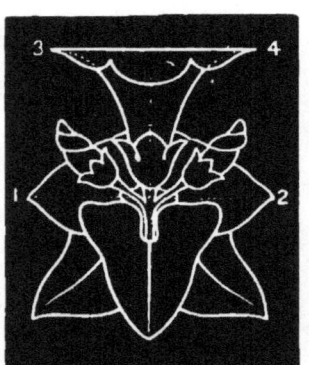

Draw a vertical line of any given length, and halve it. Make 1 2 equal to three-fourths of the vertical line, and 3 4 equal to three-fourths of 1 2. Draw the long curves from the end of 3 4 to the centre of 1 2, or very near it. On the lower third of the upper half of the vertical line, draw the calyx or cup of the flower; and, on the upper third, draw the open corolla. From the centre of 1 2, draw five leaves, also two unopened buds. See that the pupils make their drawings much larger than the copy.

Card-Exercise XIV.

Design from the Convolvulus Flower.

Draw a square, its diameters and diagonals. In the upper quarter of the square, on the semi-diameter, draw a convolvulus flower, as given in the copy, — which is the same as the flower in the last exercise, — and repeat in each of the other three quarters. Make the height of the calyx at the base of the flower about one-fourth of the semi-diameter.

It has been said that the first step in design is to take a form already drawn, and so use it as to produce a new combination. Pupils of all ages love to do this, which is an exercise partly in drawing and partly in inventing.

Do not forget to require the pupils, from time to time, to analyze the card exercises as explained in chapters two and three, before beginning to draw them. They must now describe not only the lines used, but the principles of design illustrated.

Card-Exercise XV.

Symmetrical Moresque Ornament.

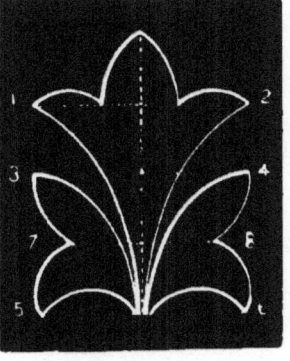

Draw a vertical line of any given length, and divide it into four equal parts. Make 1 2, 3 4, 5 6, equal to three-fourths of the vertical line, and 7 8 equal to one-half. Divide 1 2 into thirds, giving the proportions of the trefoil. First draw the long curves from 1 2; then the curves from 3 4. Finish by drawing the short, simple curves. Make all the curves as graceful as possible.

VARIATION. — Having drawn the figure as given, repeat it on the lower side of 5 6, drawing it bottom upwards. It then becomes a diaper pattern, and may be enclosed in a rectangle.

Card-Exercise XVI.
Moresque Ornament arranged as a Cross.

Draw a square. At the centre of each side of  the square, draw a line perpendicular to the side, and one-third longer. On this line, and the side of the square, draw the form given in the last card-exercise. Inside the square, draw a rosette, like *a* in card-exercise I. This illustrates how easy a thing it is to combine given units and make new designs.

Card-Exercise XVII.
Vertical Repetition. — Interlacing and Trefoil Forms.

Form *a*. — Draw the vertical line 1 2, and upon it  construct three squares, touching one another. Draw the horizontal diameters of the squares. Divide these diameters and the horizontal sides of the squares into four equal parts. Draw the oblique lines which interlace by going over and under each other alternately.

Form *b*. — Draw three squares as in the last. In each, draw a trefoil leaf, not allowing its lobes to touch either side of the square. Each of the small circles takes up one-sixth, and each of the small diamond shapes two-sixths, of a side of a square.

Card-Exercise XVIII.

Horizontal Repetition. — Trefoil Forms, and Interlacing.

Form *a*. — Draw four squares united horizontally, and add their vertical diameters. On these diameters, draw trefoil leaves, as in the copy. Draw a horizontal line above, another below.

Form *b*. — Draw four squares, united horizontally. Fill these squares with interlacing lines, as in *a*, last card-exercise. Draw a horizontal line above and below. Divide the upper horizontal line into halves; on each half construct a square, and fill it with a trefoil leaf, as in the copy. To complete the moulding, draw a horizontal line above.

Square Rosette of Conventional Leaves, and Moulding.

Draw the square, its diagonals and diameters. Make the diameter of the circle at the centre equal to one-fourth of the diameter of the square. Draw the compound curves on the diagonals. Next draw the leaves on the diameters, and finish by drawing the margins of the other leaves. Make the width of the moulding below the rosette equal to one-third of a side of the square.

VARIATION 1. — Draw three squares united horizontally, and fill them in a similar manner. Continue the moulding below the rosettes.

VARIATION 2. — Draw the square, its diagonals and diameters, and the circle at the centre. Then, instead of drawing the compound curves on the diagonals, draw them on the diameters, and the leaves on the diagonals. Finish.

CARD-EXERCISE XIX.

Toy Church.

Draw a square, and divide it into nine equal small  squares. Erase the right two of the upper three. Divide the right and left sides of the remaining upper square into thirds, and the top into fourths. Through the lower points of division on the sides, and the outer points of division on the top, draw the inclining sides of the spire. The window occupies one-half the width of the square. Draw the vertical and horizontal lines belonging to the spire and body of the church; then add the lines for the roof, door, and windows. Draw the lower lines of the roof before the upper lines; observe that the ridge is level with the bottom of the spire.

Card-Exercise XX.

Fuchsia Conventionalized.

Draw 1 2, and divide it into four equal parts. Make 3 4 equal to 1 2. Join ends of 3 4 to the upper end of 1 2 by oblique lines, and draw 5 6 through the centre of the lower fourth of 1 2. Finish the form.

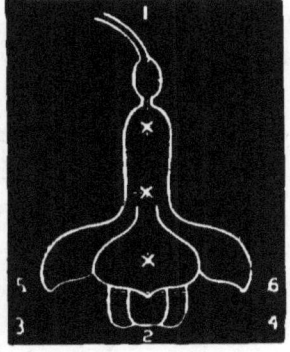

Variation. — Draw a circle, and divide it into six equal parts by drawing three diameters. On every other semi-diameter, draw a form similar to this, with the stems of the three uniting at the centre.

Card-Exercise XXI.

Pitcher, Illustrating Compound Curves.

Divide the central vertical line into four equal parts, and make 1 2, 3 4, equal to two of these parts, and 5 6 equal to three. Draw 1 3, 2 4. Finish, noting carefully where the lines of the pitcher cross the dotted lines. The width of the base of the pitcher is equal to one part and a half of the vertical line.

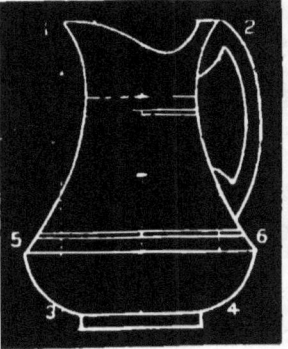

Variation. — Let the curve from 1 be unbroken at 5, rounding off to 3. Make the outline of the opposite side of the pitcher the same. Extend 1 2 half of its length beyond 2, and from its extremity draw a straight line to a point just above 6. This last line forms the outer line of the handle, to which add the thickness; then put a spout, curving above the left half of 1 2, and passing through its centre beneath the right half, but turning up again to where the handle starts.

A design must vary somewhat according to the material from which the object is to be made, as silver, clay, wood, stone, cast or wrought iron.

Card-Exercise XXII.

A Water-Bottle of Simple Curves and Straight Lines.

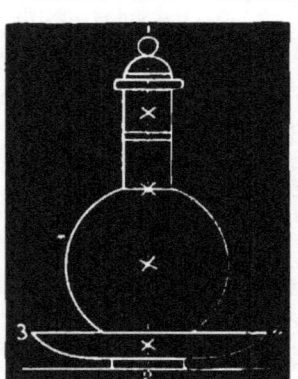

Draw the vertical line 1 2, and divide it into fourths. Just above the lower extremity, draw 3 4 equal to three-fourths of 1 2. Draw the neck and stopper, which occupy the upper half of 1 2, making the width of the neck equal to two-thirds of 1 2. On the lower half of the vertical line, draw simple curves, forming the outline of the body of the bottle, as shown in the copy. Draw the curves springing from 3 and 4. Finish.

In England the water-bottle is quite extensively used in the sleeping-rooms of hotels and in private houses. It will be seen that water put into it must of necessity be kept pure.

Candlestick, Candle, and Extinguisher.

Draw the central vertical line, and divide it into four equal parts. Through the lower point of division, draw 1 2 a little longer than three-fourths of the vertical line, and extending to the same distance on each side of it. Make 3 4 equal to one-half of 1 2. The top of the socket reaches to the centre of the vertical line, while the wick occupies the upper half of the upper fourth. Draw the curves carefully, most of them being compound. Draw the outer curve of the handle before drawing the inner. See that the sides of the extinguisher have the same slope. In this exercise but few exact measurements are given: it will, therefore, be a good one for the best pupils to draw on the blackboard from the book.

QUESTIONS. — What is said of conventionalized forms? When is it proper to imitate Nature exactly? What is mentioned as the first step in original design? How can the power of repetition in numbers on form be illustrated? How far is it proper to use the blackboard? Why should pupils in a class never draw from wall charts?

CHAPTER VIII.

PRACTICAL DESIGN.

Fruit, Compound Leaves, and Sprigs, Conventionalized.

THE pupils have now learned what is meant by conventionalized natural forms. To this manner of rendering natural forms for practical purposes, no limit can be fixed. As there are different degrees of beauty among natural objects, those objects should evidently be chosen for the purposes of ornament which will, in the first place, afford the most pleasing effect, and, in the second place, will be best adapted to the particular purpose for which they have been selected. In one instance it will be a flower, in another a sprig, in another a vine, and again it will be something else. Thus a design for a wall paper, to be viewed only from one position, may be made quite different from a design for a carpet, to be viewed from all sides. The view should always be agreeable, but cannot be if the design is seen inverted. Position, therefore, is to be considered; the floor and ceiling, for example, are to be distinguished from the walls of the room. Regard must also be paid to shape, and the ornament be modified for brackets, corner-pieces, finials, corbels, mouldings, columns. Again, no design is good that violates the sense of fitness; construction must never be adapted to ornament, but always ornament to construction, — to the object and its use.

Card-Exercise XXIII.

Acorn and Cups Conventionalized.

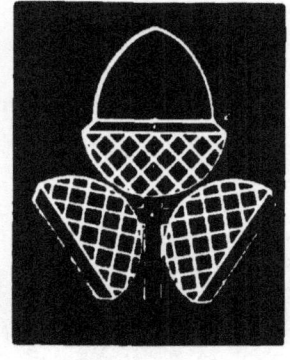

Draw a vertical line, and divide it into three equal parts. Through the upper point of division, draw 1 2 slightly longer than one-third of the vertical line. Above this, draw the outline of the acorn; below it, draw its cup, descending nearly to the second point of division on the vertical line. The outline of the cup is a semi-circle. Add the empty cups, making them like the full cup.

VARIATIONS. — Draw the figure as given in the copy; then fill the side cups with acorns of the same shape as the acorn in the upper cup.

Card-Exercise XXIV.

Acorns and Leaves Conventionalized.

Draw the acorns as in the last exercise, and then continue the central vertical line downwards until its length is doubled. Draw two oak-leaves springing symmetrically upwards from the bottom of the vertical line, and touching the sides of the acorns. There are different varieties of the oak-tree, and of course they have leaves somewhat different in form If it can be done, request the pupils to bring in some oak-leaves, also acorns, and compare them with the copy.

Card-Exercise XXV.

Clover-Leaf Conventionalized.

Draw a vertical line and halve it. On the upper half, as a midrib, draw the top lobe. From a point just below the base of the top lobe, draw the midribs of the lower lobes. Complete the lobes.

VARIATIONS. — Draw an equilateral triangle, and from its centre draw a straight line to each angle. On these lines as midribs, draw a compound leaf like the one in the copy.

Draw two straight lines bisecting each other at right angles. Divide each into thirds, and, through the points of division nearest the common centre of the two lines, draw a circle. On each of the outer parts of the lines, draw a single lobe of the clover-leaf, thus obtaining the effect of the four-leaved shamrock.

Again, draw two equal lines, one horizontal, the other vertical, bisecting each other. Divide each into four equal parts. On the upper half of the vertical line, draw a complete clover leaf pointing upwards; on the lower half, draw a second pointing downwards.

On each end of the horizontal line, draw a single lobe of the leaf. The result will be a quatre-foil.

Card-Exercise XXVI.

A Conventionalized Compound Leaf.

Draw the vertical line forming the rachis of the 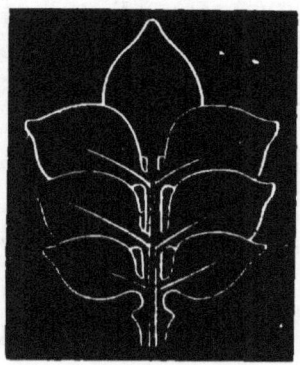 leaf, and the oblique midribs of the leaflets. Then draw the outline of each leaflet.

When the leaflets of a compound leaf are smooth-edged, draw each complete, as shown by the dotted parts in the copy. Afterwards erase the parts which are not visible. Visible parts will thus be made consistent with one another, which is not often the case unless the whole form is drawn. It is also the most expeditious method, requiring the fewest corrections.

VARIATIONS. — With the same proportions, rachis, and midribs, vary the figure by giving to each leaflet three lobes, as in the trefoil leaf. The top leaflet may be omitted, and a flower put in its place. The potato flower will suit the leaf best.

Again, draw two straight lines bisecting each other at right angles. On this cross, draw four compound leaves, like the one given in the copy. Between each pair, draw a cinque-foil flower, that is, a flower having five leaves; or draw a bunch of spheres which shall represent the ball or seed of the flower.

Card-Exercise XXVII.

Wine-Glass.

Having divided the vertical line into four equal parts, make 1 2, 3 4, equal to two of these parts, 5 6 to one of the parts, and 7 8 a little shorter than 5 6. Draw in this order: bowl, stem, base.

VARIATIONS. — With the same proportional division of the vertical line a great variety of shapes may be obtained. Thus, make 1 5, 2 6, straight, and the stem plain.

Again, having drawn 1 5, 2 6, straight, use them as construction lines, halving them, and through the points of division drawing compound curves to form the profile of the bowl. In the centre of the stem, draw a circle with a diameter one-eighth of the vertical line. Base of glass the same in the three cases. Draw parallel horizontal double lines near the lip, and between them draw a star repeated around the glass. The star may be composed of four straight lines crossing one another at a common centre, the alternate lines mutually perpendicular.

See that the pupils, when they examine their drawings, do not look at them obliquely. They should hold them directly in front of the eye, and turn them around. Unless frequently reminded, they will neglect to do this.

Card-Exercise XXVIII.

A Moulding of Heart-shaped Leaves.

Draw a square and its vertical diameter. Through a point a little below the centre of this diameter, draw a horizontal line indicating the greatest breadth of the leaf. Draw the compound curves as in the copy. Next, divide the lower side of the square into four equal parts; on each part, construct a square, and fill it as the large square was filled.

Card-Exercise XXIX.

A Mug.

Draw the central vertical line, and divide it into six equal parts. Make 1 2, 3 4, equal to two-thirds of the vertical line, and draw the sides of the mug. Draw the horizontal lines, the lower one of the upper group through the upper point of division on the vertical line, the upper one of the lower group through the lower point of division, and the single one through the central point of division. Observe that the flutings towards the bottom decrease in width to right and left, because they are seen ob-

liquely. Make the handle at its widest point equal to one-third of the vertical line, with its greatest projection from the mug towards the top. Notice the horizontal lines from which the handle springs. Draw last the inner line, which gives the thickness of the handle. Pupils incline to make the handle thick and clumsy. Anticipate and prevent this.

VARIATION. — Divide the central vertical line into quarters, and make the width of the mug equal to half its height. Keep the other details the same, but make the handle half square in shape, projecting as far as in the copy, and equally at top and bottom.

CARD-EXERCISE XXX.

Maple-Leaf Conventionalized.

Draw a vertical line, and divide it into thirds. Make

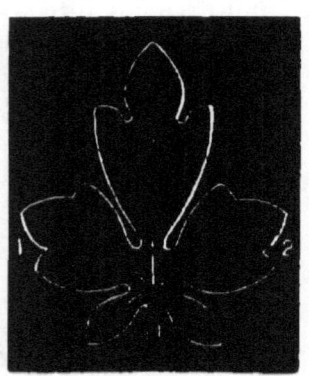

1 2 equal to two-thirds of the vertical line. Divide 1 2 into thirds, and, through the points of division, draw the veins of the side lobes. Complete the leaf as shown. Notice that the curves which enclose the lobes of the leaf always bend outwards from the centre of the lobe; also that the lines at the junction of the lobes nowhere form angles, but have a rounded shape. Require the pupils to draw this leaf from time to time, until they can readily draw it from memory.

Card-Exercise XXXI.

Maple-Leaf Design.

Draw a horizontal and a vertical line of the same given length, bisecting each other. Divide each into thirds. Unite the points of division, thus forming a square on its diagonals. Within the square, draw a trefoil; within the trefoil, a curved equilateral triangle. On the outer third of the vertical and horizontal lines, draw conventionalized maple-leaves, having the proportions and shape of the one in the last card-exercise.

Card-Exercise XXXII.

Conventionalized Form; Simple Leaves and Flowers.

Draw the vertical line 1 2, and divide it into four equal parts. Make 3 4, 5 6, equal to one-half of the vertical line. 3 4 to be drawn through the upper point of division on the vertical line. Draw first the midribs and stems of the leaves, and then their margins. Draw the flower at the top, and lastly the little circles at the bottom, representing the berries.

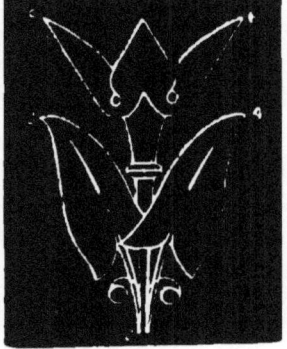

Card-Exercise XXXIII.

Conventionalized Form.   Trefoil Leaves and Flowers.

Draw a vertical line of any given length, and divide it into thirds. Make 1 2 and 3 4 equal to, and 5 6 a little less than, two-thirds of the vertical line. Connect 3 5 and 4 6. From the ends of 3 4, draw the long curves, giving the centres of the two leaves and stems. From the ends of 1 2, draw the compound curves which form the outline of the flowers. These curves, after touching 3 4 at a short distance from the vertical line, turn upwards, and unite on the vertical line one-third of the way from 3 4 to 1 2. Draw the stem of the flower from the points where its outline touches 3 4. Add, by measurement of eye, the details of the flower, leaves, and pendent berries.

QUESTIONS. — What is said of the limit to conventionalized natural forms? By what are you to be guided in selecting different forms for the purposes of ornamentation? How should ornament be influenced by position? By shape? By use?

CHAPTER IX.

FORMS OF LEAVES AND FLOWERS.

STUDY OF BOTANY.

It has already been said, that Nature often builds upon marked geometrical forms. In details she is irregular, but usually the reverse in general outlines. That this is true of leaves and flowers, will be seen from the exercises which immediately follow. Among leaves, some have the general form of the ellipse, others of the oval, others of the triangle, while others are modelled upon the pentagon. Among flowers, there is a less variety of forms, since their petals are usually arranged symmetrically around a centre.

Request each pupil to bring to school the leaf of any designated tree, shrub, flower; as the maple, rose, lily. Each pupil having in his hand the same variety of leaf, proceed to give a lesson on its form and structure, teaching at the same time both botany and the principles of practical design. For a second lesson, take the same leaf, but add another, and teach the pupils to compare the two. Thus proceed from one leaf and flower to another, and at last require each pupil to bring to school the largest possible number of different leaves, and again of flowers. This work should, of course, be done at the most favorable season of the year.

Acquaint the pupils with the proper botanical

terms. The following are some of the terms it will be necessary to use; others can be readily obtained from any good work on botany, if you are not already master of the subject, as every teacher of children should be.

PETIOLE. — This is the leaf-stem, or leaf-stalk.

MIDRIB. — This is the continuation of the petiole, and forms the axis of the leaf.

RIBS. — When the petiole divides, as it does in the maple-leaf, each of the divisions is called a rib. There is then no midrib.

VEINS. — These are the small branches which put out from the midrib, or from the ribs.

VEINLETS. — These are small branches putting out from the veins.

BLADE. — This is the broad, flat part of the leaf.

VENATION. — The lines seen upon the blade constitute what is called its venation.

MARGIN. — This is the edge of the leaf. It may be entire, that is, unbroken, as with the lilac; it may be serrate, having sharp notches, as with the syringa; it may be crenate, having broad, round notches, as with the catnip-leaf. There are many other varieties.

A LOBED LEAF. — When the notches in the margin of a leaf are deep, as with the oak and maple, the leaf is said to be lobed. The deep notches are called sinuses.

BASE AND APEX. — The part of the blade next the petiole is called the base of the leaf; the opposite part is called the apex. These are of all shapes, as cordate, or heart-shaped; hastate, or spear-shaped; acute, or sharp; obtuse, or blunt; and so on.

COMPOUND LEAF. — This consists of several leaves, or rather leaflets, attached to a common rachis, as in the rose-leaf.

RACHIS — This is the continuation of the petiole, or stalk, to which the leaflets of a compound leaf, or to which the stems of a cluster of flowers, are attached.

PEDUNCLE. — This is the flower-stalk of a single flower or the rachis of a cluster of flowers.

FORMS OF LEAVES AND FLOWERS.

CALYX. — This is the cup, or outer circle, of leaves which belong to a flower.

SEPAL. — This is one of the leaves composing the calyx.

COROLLA. — This consists of the usually delicately colored flower-leaves inside the calyx.

PETALS. — These are leaves of the corolla.

STAMENS. — These are the slender parts found next inside the corolla.

PISTIL. — This is the central part of the flower within the stamens.

FLOWER IN PLAN AND ELEVATION. — The first is the flower as seen when viewed directly from above; the second as seen when it is viewed from one side.

CARD-EXERCISE XXXIV.

Syringa-Leaf slightly Conventionalized.

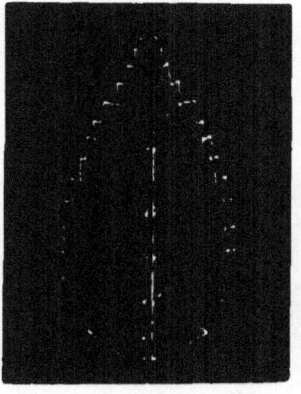

Draw a vertical line forming the midrib of the leaf, and divide it into five equal parts. Through the second point of division from the bottom, draw a horizontal line one-half as long as the vertical line. Add the veins joining the midrib. Around the whole, draw the serrate margin of the leaf. The apex of the leaf is acuminate; the base obtuse. The whole leaf is slightly conventionalized, making it symmetrical. As a whole, the leaf is oval, inclining to the ellipse.

With the proportions thus given, the pupils can readily draw the leaf many times larger than in the copy.

Card-Exercise XXXV.

Catnip-Leaf, slightly Conventionalized.

Draw a vertical line forming the midrib and petiole 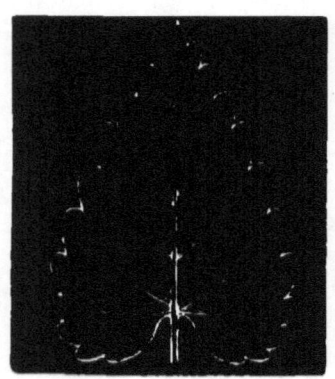 of the leaf, and divide it into thirds. Through the lower point of division, draw a horizontal line two-thirds as long as the vertical line. Add the veins. Around the whole, draw the crenate margin of the leaf. The base of the leaf is cordate, or heart-shaped; its apex acute. The petiole, or leaf-stalk, occupies one-half of the lower third of the vertical line. As a whole, the leaf is decidedly oval.

Card-Exercise XXXVI.

German Ivy-Leaf, slightly Conventionalized.

Draw a vertical line forming the midrib and petiole  of the leaf, and divide it into three equal parts. Through the points of division and through the lower end of the line, draw horizontal lines, the upper and lower a little more than half, the central a little less than half, as long as the vertical line. Add the side ribs, and, around the whole, draw the margin of the leaf. The margin is entire; the leaf five-lobed, with sinuses very shallow. As a whole the leaf is pentagonal.

Card-Exercise XXXVII.

Gooseberry Leaf, slightly Conventionalized.

Draw a vertical line forming the midrib and petiole of the leaf, and divide it into thirds. Just below the upper point of division, draw a horizontal line not quite as long as the vertical line. Add the other ribs and the veins, and around the whole draw the margin of the leaf. The leaf is five-lobed, sinuses deep, margin unevenly serrate.

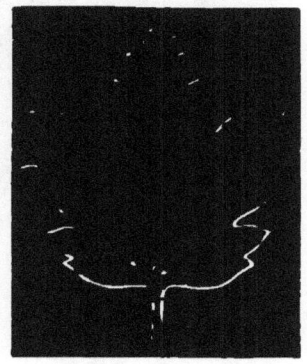

Card-Exercise XXXVIII.

Strawberry-Leaf, slightly Conventionalized.

Draw a vertical line forming the petiole of the leaf and the midrib of the central leaflet. Divide this line into thirds, and, through the upper point of division, draw a horizontal line as long as the vertical line. Draw the midribs of the right and left leaflets, and add the veins for all three. Finish by drawing the coarsely serrate margin of each leaflet.

A Leaf, slightly Conventionalized.

Draw the vertical line, forming the midrib, and divide it into thirds. Just above the centre, draw a horizontal line, making it equal to two-thirds of the vertical line. Add the other horizontal lines, one below and two above, by judgment of eye. Then draw the lines forming the veins, and, around their ends, draw the margin of the leaf.

This leaf is just enough conventionalized to make it symmetrical. As it is a simple leaf, the midrib is the axis.

Currant Leaf, slightly Conventionalized.

Draw the vertical line, forming the midrib of the leaf, and divide it into four equal parts. Through the upper point of division, draw the horizontal line, making it equal to three-fourths of the vertical line; through the lower point of division, draw a second horizontal line, making it a little shorter than the first. Add the lines forming the veins, and, around their ends, draw the margin of the pentagonal leaf.

This leaf is, also, just enough conventionalized to make it symmetrical on its midrib, which is the axis.

Hawthorn-Leaf, slightly Conventionalized.

Draw the vertical line, forming the midrib, and divide it into four equal parts. Through the central point of division, draw the upper horizontal line, making it a little longer than one-half of the vertical line; and, through the lower point of division, draw a second horizontal line, making it a little shorter than three-fourths of the vertical line,  but longer than the first horizontal line. Add the lines forming the veins, and, around their ends, draw the margin of the leaf, as shown in the copy.

This leaf is not quite from nature, but is conventionalized just enough to make it symmetrical on its midrib. The left side balances the right.

Geranium-Leaf, slightly Conventionalized.

Draw a vertical line. From a point just below the centre, draw the lines forming the ribs; and, around these, draw the contour of the leaf, making the breadth a trifle greater than its height.

Use these different leaves not on the cards, for blackboard lessons, and for the purpose of giving further illustrations of principles of design,  and for imparting a knowledge of botany.

Card-Exercise XXXIX.

Laburnum-Leaf, slightly Conventionalized.

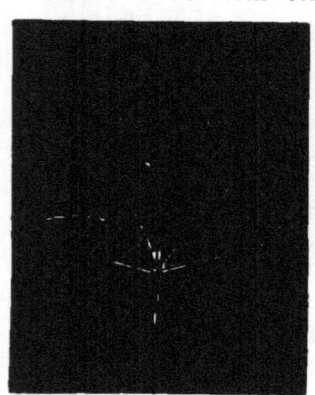

Draw a vertical line, forming the petiole of the leaf and the midrib of the central leaflet. Divide this line into four equal parts. Make the midribs of the right and left leaflets about two-thirds as long as the midribs of the central leaflet. The breadth of each leaflet is a little less than one-half of its length. The margin is entire. This is a compound leaf, each leaflet elliptical; general outline of the whole triangular.

Card-Exercise XL.

Rose-Leaf, slightly Conventionalized.

Draw a vertical line for the rachis, and mark the points of junction made by the leaflets with it. Draw the midribs of the leaflets, and finish. Each leaflet is elliptical: the whole leaf, in its general outline, might be called the same. This is a compound leaf, which is quite a different thing from a lobed leaf. Sometimes, however, it is no easy matter to distinguish one from the other.

Card-Exercise XLI.

Wild-Rose Leaves and Flowers from Nature.

At *a* we have the plan of the flower, or the flower as seen when we look down upon it. This is pentagonal, cinque-foil, or five-leaved. At *b* we have the elevation of the flower, or the side view. At *c* we have the bud; and at *d* the compound leaf. It will be seen, that the margin of the natural leaflet is serrate, and not entire, as given in the last exercise, where the leaf is slightly conventionalized.

Observe how symmetrically the five large petals of the corolla at *a*, also the five small petals of the calyx, making ten in all, are arranged about the centre of the flower. The ten little stamens near the centre have the same symmetrical arrangement. Observe, also, how symmetrically the leaflets of the compound leaf at *d* are arranged on an axis, that is, on the rachis from which the leaflets spring.

As the pupils draw the leaves and flowers, question them about their forms and parts. Do this until they are perfectly familiar with the terms used. At the same time question them, also, about the principles of design to be learned from these natural forms. You can readily make this a very interesting exercise. In the same way shells and insects may be drawn, and lessons in design learned from them.

Card-Exercise XLII.

Crane's-Bill Flower, Leaf, and Bud from Nature.

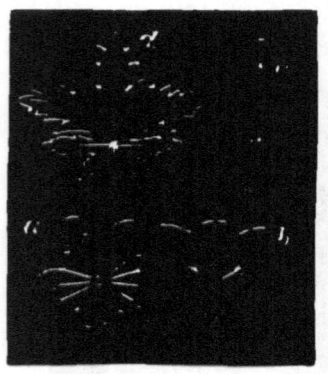

At *a* we have the plan of the flower; at *b* the elevation; at *d* the leaf; at *c* the bud, whose shape gives the name, — crane's bill. The flower is pentagonal, also the leaf. Take notice of the symmetrical arrangement of petals and stamens about the centre of the flower at *a*. Compare the leaf at *d* with the leaf at *d* in the last card-exercise. This is deeply lobed and simple, while the other is compound. This is symmetrically balanced on its central rib; that is, on its rachis.

For at least a part of each year, botany should be taught in every primary school; and to a certain extent it should be taught in connection with drawing and designing. The reproduction with the pencil of the forms of leaves, buds, flowers, sprigs, vines, ferns, etc., would not only be good practice in drawing, even if the forms were only conventionalized, all the little details being omitted, but would quicken the perceptive powers much more than the mere act of observing the forms as a student of science, and thus would firmly fix the botanical knowledge in the mind. And then, the study of the manner in which Nature, the great architect and decorator, uses her material, distributes and strengthens her lines, covers and beautifies surfaces, would reveal the laws that must be obeyed in practical design, if we would secure the happiest effects. Thus, in a certain degree, to unite botany, drawing, and designing, would not only please and profit all, even young children, but it is an easy thing to do when there is a will to do it.

Begin with a leaf, any leaf, the pupils being supplied with newly-gathered ones. This is among the first questions to be asked: What is the general form of the leaf? That of the circle, ellipse, oval, triangle, pentagon? Then would follow the minor distinctions of cordiform, deltoid, hastate, etc. What is the character of the margin? Is it entire, serrate, repand, lobed, or something else? How are the ribs and veins distributed, and do they unite tangentially, or otherwise? As you proceed, there will be such questions as these: What lessons in symmetrical arrangement about a centre, along an axis, are to be learned from simple leaves, compound leaves, from the petals and stamens of flowers? What lessons in radiation from a parent stem are to be learned from deeply-lobed and compound leaves and from flower-umbels? What lessons in continuous development for covering irregular and other surfaces can vines, trailing, climbing, twining, and axillary inflorescence, teach us? What lessons in breadth, in repose, can we learn from all? Nature has a ready response for each of these questions, whenever we choose to interrogate her in a proper manner. But, while she is answering the questions put by the student of art, she can, at the same time, answer many of the questions put by the student of science.

The exercises on the next page are given for teachers who have acquired some skill in drawing, to put on the board rapidly before their classes. They are given here merely as suggestive of a wide variety of exercises which teachers may use to make their drawing-lessons interesting to their pupils. The drawings of the worm, the chrysalis, and the butterflies, are simple illustrations of how the drawing-exercises may be made the means of teaching much valuable knowledge, as well as developing the pupils' perceptive faculties. What an interesting lesson in natural history can be made from these simple forms, even though they be but roughly executed! The author of this little work looks forward confidently to the time when drawing in public schools shall not be considered merely as an exercise by itself, but when teachers shall use their skill in drawing largely to illustrate and teach many other branches of study.

QUESTIONS. — What is said of the use of geometrical forms by Nature? Repeat the botanical terms given. What is said about questioning pupils in their drawing of flowers, leaves, &c.? What is said about the teaching of botany in primary schools? How can the drawing-exercises be made interesting? How can skill in drawing be employed in teaching?

A SYSTEM

OF

Industrial and Artistic Drawing

FOR PUBLIC SCHOOLS.

PREPARED BY PROF. WALTER SMITH,

State Director of Art Education for Massachusetts, General Supervisor of Drawing in the Boston Public Schools, and Director of the Massachusetts Normal Art School.

This system is the only comprehensive course of instruction in drawing accessible to American schools. The course is so graded as to meet the wants of every class of pupils, from the lowest primary class to the most advanced class of the high school.

The system comprises

A Primary Course,
An Intermediate Course,
A Grammar Course,
A High School Course.

THE PRIMARY COURSE.

This course consists of a Manual for the use of teachers, in which the simple elements of the study are explained and illustrated by the most familiar terms and examples; and two series of Cards, containing exercises for pupils to draw on their slates.

Price of the Manual, $1.00; of the Cards, 15 Cents each set.

THE INTERMEDIATE COURSE.

This course consists of three small Drawing-Books, of twenty pages each, specially arranged for pupils when they begin to draw on paper. The exercises are of a similar character to the card-exercises, with the addition of some simple exercises in freehand Perspective.

This course also contains a Manual for teachers.

Price of the Manual, $1.00; of the Books, 15 Cents each.

THE GRAMMAR COURSE.

This course consists of : —

First. FOUR BOOKS IN FREEHAND OUTLINE DRAWING AND DESIGN. The exercises in these books are more advanced than those in the Intermediate Course; and by a wide variety of ornamental, conventional, and natural forms, and representations of historical ornament, pupils are taught a great deal about the art of past ages, and also the principles of good Design.

Second. FOUR BOOKS IN GEOMETRICAL DRAWING. These books form the basis for Perspective, Model and Object, and Mechanical Drawing. The exercises consist of problems in Plane Geometry, the working of which teaches pupils the exact meaning of words and terms; and by the care required to execute the problems they are trained to accuracy of workmanship.

Third. TWO BOOKS IN MODEL AND OBJECT DRAWING. The exercises in these books are all in outline; and pupils are taught in a thorough manner how to draw from objects. The exercises are of such a character, that the pupil's taste will be cultivated while acquiring skill in drawing.

Fourth. TWO BOOKS IN PERSPECTIVE DRAWING. These books teach the elements of Parallel and Angular Perspective.

Each of these divisions is accompanied by a Manual for teachers, containing all the exercises in the books, and many more besides.

Price of the Books, 25 Cents each.

As above set forth, the Grammar Course comprises twelve books; and with an allowance of an hour and a half to two hours per week to drawing, pupils can easily go through three of these books in a year; and when pupils have finished this Grammar Course, they will be able to draw in outline whatever they can see and understand.

Where pupils in grammar schools have received no previous instruction in Freehand Drawing, the first grading of the instruction must be provisional.

THE HIGH SCHOOL COURSE.

This course consists of advanced work in Perspective, Model and Object Drawing, or Mechanical Drawing, according to the tastes of pupils.

As in the previous courses, only outline work has been attempted, in order that pupils might become well grounded in all the elementary principles of Industrial or Artistic Drawing, when they reach the high school, they will be able to take up understandingly the more advanced phases of the study. In the high school, pupils may be allowed some election in their course of study, according to their tastes or inclinations: some may prefer a purely artistic course, and others a mechanical course. Their previous training fits them for either course.

Books for this course are in preparation.

The High School Course includes instruction in Shading, Painting, Drawing from Nature, and Designing in Color.

THIS SYSTEM IS A PRACTICABLE ONE.

It is frequently represented that this system is too elaborate for public schools. It may be said, in answer to this, that as shading and figure-drawing, &c., are placed at the end of the course, and outline forms are used to give the elementary instruction, ample time is secured during the eight to twelve years of school-life, to impart a sound knowledge of Drawing.

This system proceeds on the assumption that mere hand-skill in Drawing is of secondary importance; that Drawing, like writing, should be used principally as a means of expression.

Starting on this general principle, it is only necessary that the knowledge given during the period of a pupil's school-life should be such as he can comprehend in the course of his advancement. The expression of this knowledge by drawing will become by practice of no more difficulty than expressing thoughts by writing; and if pupils draw during their whole school-course, they will learn to draw well as readily as they will learn to read or write well.

THIS SYSTEM CAN BE TAUGHT BY REGULAR TEACHERS.

The teachers' manuals which accompany the books give full particulars in regard to teaching each subject; and any teacher, with a little patient study, can fit himself to teach understandingly and well what is required to be taught below the High School Course.

This system has the indorsement of some of the leading educators of the country, and is rapidly being introduced into public schools. Although the books have been published but a few months, the system has been adopted by the Boards of Education of the following cities:—

Boston, Lawrence, Lowell, Worcester, Gloucester, Fitchburg, Taunton, New Bedford, Newton, Waltham, Fall River, Dedham, Woburn, Arlington, Concord, N.H., Saratoga Springs, Syracuse, Pittsburgh, Columbus, Detroit, Indianapolis, Milwaukee, Oakland, Cal., Sacramento, Cal.; and the Primary Course has been adopted for the Public Schools of the State of California.

www.ingramcontent.com/pod-product-compliance
Lightning Source LLC
Chambersburg PA
CBHW020300170426
43202CB00008B/445